SCIENCE TEXTBOOK CONTROVERSIES AND THE POLITICS OF EQUAL TIME

SCIENCE TEXTBOOK CONTROVERSIES AND THE POLITICS OF EQUAL TIME

Dorothy Nelkin

The MIT Press
Cambridge, Massachusetts, and London, England

This book was set in V-I-P Baskerville by The MIT Press Computer Composition
Group, and printed and bound by The Colonial Press Inc. in the United States
of America.

Library of Congress Cataloging in Publication Data

Nelkin, Dorothy.
    Science textbook controversies and the politics of equal time.

    Includes index.
    1. Science—Study and teaching—United States.
2. Science—Text-books. I. Title.
Q183.3.A1N44   507'.1073   76–58459
ISBN 0–262–14027–6

To Laurie, Lisa, and Alan

Tables   viii

Abbreviations   ix

Preface   x

1 Introduction   1

I THE CONTEXT
2 A Science or a World View? Historical Notes on the Teaching of Evolution   9
Evolution and the Nineteenth Century Soul   9
Antievolution in the 1920s   13
The Legacy of the Scopes Trial   16
3 The Science Curriculum Reform Movement   21
The Federal Government's Role   22
Two New Courses   27

II THE SCIENCE TEXTBOOK WATCHERS
4 Textbook Watchers and Space Age Fundamentalism   41
Forbidden Subjects   41
The Conservative Ministries   43
Some Activists and Organizations   47
Political Tactics   50
5 The Scientific Creationists   60
The Bible as Science   61
Creationist Organizations   65
The Activists   71

III TEXTBOOK DISPUTES
6 Creation vs. Evolution: The California Controversy   81
Creationist Demands   82
Evolutionist Response   85
The California Solution   94
7 The Proper Study of Mankind. . . .: The MACOS Dispute   104
"We're all animals, kids are taught here"   105
The Politics of Local Protest   107
MACOS: A National Debate   111
The Tip of the Iceberg: MACOS, Accountability, and the NSF   117

IV SCIENCE AND THE RESISTANCE IDEOLOGY
8 Social Sources of Textbook Disputes   127
Disillusion with Science and Technoloy   128
Challenges to Authority   131
The Ideology of Equal Time   134
9 Science and Personal Beliefs   144
Images of Science   144
Problems in the Communication of Science   148

APPENDIXES
1. NSF Precollege Science Curriculum Project Grants (1957-1975)   157
2. Public Knowledge of Science: Report of a Survey by the National Assessment of Educational Progress   164
3. Proposed Creationist Revisions of the California *Science Framework* for 1976   167

Index   171

Tables

1. NSF Support of Four Precollege Projects for Education in the Sciences (1958–1974). *(25)*

2. NSF Grant Funds Awarded for Development, Evaluation, and Implementation of MACOS Curriculum (1963–1975). *(35)*

3. Membership Statistics for Conservative Churches (1960 and 1970). *(44)*

4. Changing Church Membership in Twelve Protestant Groups (1958–1974). *(45)*

5. Alternative Models: Creation vs. Evolution. *(63)*

6. Advisory Board and Staff of the Institute for Creation Research (1973). *(74)*

7. Contrasting Arguments of Creationists and Evolutionists. *(89)*

8. Changes in Biology Textbooks Recommended by the California Board of Education. *(98)*

9. Changes in Average IQ Scores of a Sixth-Grade Class During a MACOS Dispute. *(107)*

10. Man: A Course of Study. Domestic Sales (1971–1975). *(108)*

11. Parallel Concerns Among Diverse Critics of Science and Technology. *(139)*

Abbreviations

AIBS American Institute of Biological Science

ASA American Scientific Affiliation

CBE Council for Basic Education

BSCS Biological Sciences Curriculum Study

CDA Curriculum Development Associates

CRS Creation Research Society

CSRC Creation Science Research Center

EDC Education Development Center

ESI Educational Services Institute

ICR Institute for Creation Research

MACOS Man: A Course of Study

NABT National Association of Biology Teachers

NAEP National Assessment of Educational Progress

NSF National Science Foundation

PSSC Physical Science Study Committee

Preface

This study began simply out of curiosity about the creationists as a
group of people who dared to represent themselves as scientists
while challenging the most sacred assumptions and norms of the scientific establishment. As the disputes developed over Man: A
Course of Study (MACOS), a National Science Foundation course in
the social sciences, the creationists' demands, which had seemed so
bizarre, began to appear to be an expression of basic and rather
widespread criticism of science and its pervasive influence on social
values. Thus, the study turned to analyze the creationists and anti-
MACOS disputes as a manifestation of some of the more abstract and
vague concerns about science and technology. Science is attacked
these days from both the left and the right. The disputes described
here represent a reaction from a highly conservative population; but
interest in local control and increased participation and concern
with the dominance of scientific values and the role of expertise will
be familiar to people located anywhere along the spectrum of political ideologies. The textbook disputes provide a means of exploring
the relationships between science and its public and of examining
how the growing criticism of science can bear on public policy.

The research for this study was facilitated by the cooperation of
many groups. These included scientists, creationists, teachers, educational administrators, parents of school children affected by the
disputes, and book publishers. People from the following organizations provided material from their files and cooperated during extended interviews: The National Association of Biology Teachers,
the Biological Sciences Curriculum Study, Education Development
Corporation, Curriculum Development Associates, the National Science Foundation, the National Association of Educational Progress,
the National Education Association, the Council for Basic Education, the Institute for Creation Research, Education Research Analysts, the American Scientific Affiliation, and the Creation Science
Research Center. In addition, biologists, school teachers, fundamentalist ministers, politicians, congressional aides, and individuals

involved in controversies throughout the country cooperated in interviews, providing useful documentation on their perspectives. As in any controversial situation, a number of those who were interviewed may disagree with my analysis, but I very much appreciate their cooperation and have attempted to use their material fairly.

I also appreciate the criticism received from scholarly colleagues who have spent considerable time reviewing drafts of this manuscript and some of the articles and seminars developed from it; these include William Blanpied, Harvey Brooks, Gerald Holton, Everett Mendelsohn, Larry Moore, Mark Nelkin, Gerard Piel, and Will Provine. Articles from the manuscript have appeared in Gerald Holton and William Blanpied, *Science and Its Public* (G. Reidel, 1976), and in *Scientific American*, April 1976. Editorial criticism in preparing these articles and the many letters that followed their publication were extremely helpful.

Support for this study was provided in part by a National Science Foundation research grant on science policy. Some of the writing was done during a semester at the Political Science Department and the Center for International Studies at MIT. I would like to acknowledge their hospitality and invaluable assistance. Finally, I appreciate the support of my colleagues at the Cornell University Program on Science, Technology, and Society and Department of City and Regional Planning.

1 Introduction

One day in 1969, the people of California learned that their state board of education had issued new guidelines for its public school biology curriculum. These guidelines included a statement that the book of Genesis presents a viable scientific explanation of the origin of life and that creation theory should be taught in biology courses as an alternative to the theory of evolution. It was only fair, it was claimed, that equal time be given to the two theories and that students be allowed to take their choice.

The California guidelines were a dramatic manifestation of an intense and persistent conflict over the values conveyed to students through the teaching of biology and the social sciences. School boards and curriculum committees in many communities have been effectively prevented from recommending biology and social science textbooks that appear to collide with traditional beliefs. In 1972, the issue reached the courts, when the religion editor of the *Washington Evening Star*, who claimed to be acting in the interest of forty million evangelistic Christians in the United States, sued the National Science Foundation for using tax monies to support education that violated religious beliefs. It reached the United States Congress in 1975 following an epidemic of community controversies over the NSF-sponsored social science curriculum called Man: A Course of Study (MACOS), a course based on evolutionary assumptions concerning the relationship of man to animals and the adaptation of man to his physical environment. Such controversies have forced a major re-evaluation of the massive federal effort to modernize the precollege science curriculum that began after the launching of Sputnik in 1957.

The textbook disputes are one expression of a renewed concern with moral and religious values in American society and a related ambivalence toward science, often perceived as a dominant world view that threatens traditional values and suppresses essential elements of human experience. Indeed, at a time when the accomplishments of science have fostered faith in the value of rational explanations of nature, there are concentrated efforts to reinvest the educational system with traditional faith. At a time when the evolutionary concepts of biology are among the more firmly established generalizations of modern science, public demands for the teaching of alternative explanations of human origins have exercised remarkable influence. At a time when the complexity of the modern technological enterprise requires specialized expertise, there are increasing demands for lay participation in the selection of science curricula by community groups insisting that education must reflect community values. These are some of the paradoxes manifest in a bitter dispute over science textbooks that has revived issues assumed to have expired with the growth of confidence in science following World War II.

The science textbook watchers have targeted courses that were developed in the 1960s to bring the concepts and methods of modern scientific research to the nation's public schools. NSF has funded fifty-three science curriculum projects, bringing university scholars together with professional educators to create textbooks and other course materials. For years this effort was welcomed as a major and important reform, "a victory of human reason over obscurantism," a "renaissance movement," in education. But in the late 1960s, in a society increasingly concerned with the social implications of science, rather suddenly university scholars and professional educators became an "arrogant elite," imposing their values on the American school child. Critics saw scientific theories invested with social and moral persuasion, threatening traditional values and personal beliefs. The NSF, once a source of intellectual enlightenment, was accused of promoting "value-laden" materials, and was pressured to review the courses produced under its sponsorship and to increase lay participation in course evaluation.

Why, one must ask, in this science-dominated age, is there such concern with science education? Why has the old resistance to evolution theory gathered new momentum? What issues have converged to force public recognition of complaints long ignored as merely the rumblings of marginal groups of religious fundamentalists and right-wing conservatives? How have small groups of "believers" been able to intrude their ideologies into educational establishments and in some cases to control the educational apparatus that determines science curriculum?

Public school education is among the most volatile areas of public policy, and efforts at reform can count on opposition. Conflict thrives on the persistent concern of parents about what influences their offspring and on considerable public ambivalence about the role of education. Should schools emphasize individual intellectual development or should they be responsible for the transmission of culture? Should they convey basic knowledge or provide useful skills? Are they to generate new values or preserve existing ones? Should they promote social change or protect the status quo?

The history of American educational reform has reflected an abiding faith that schools are a means to remedy social problems and to bring about social reform. Education is often viewed as an ideological instrument, a means of changing social perceptions, such as racial or sexual prejudices. The perceived importance of education as an instrument of reform is also the basis of a powerful conservative reaction among those who seek to protect traditional ideologies. Curriculum reform often becomes the focus of bitter disputes over the values to be conveyed to the young.

Yet, with the notable exception of the Scopes trial (often thought to be the last vestige of the great struggle between religion and science), there have been relatively few public attacks on science teaching. Following World War II, science, perceived as neutral and associated with material progress, was dissociated from value questions of interest to textbook watchers. The pervasive influence of science on advanced technological societies suggested that traditional religious beliefs no longer had significant bearing on science education policy.

In the late 1960s, however, attitudes appeared to change. Growing concern with the misuse of science and the harmful effects of technology led also to criticism of scientific rationality. The interest in nonrational and supernatural explanations of nature is reflected in the proliferation of cults and sects based on Eastern mysticism. Less visible, but perhaps more important in the long run, has been a growth in fundamentalist churches, especially in the very centers of high technology industry, such as urban Texas and southern California. In these areas, many conservative citizens expressed disillusion with the "decadence of scientism" and the bureaucratic authority that seemed to remove their sense of local power. The dominance of scientific values and the consequent decline in Christian teaching, they claimed, were responsible for contemporary social ills. These people formed the financial, social, and political base for the science textbook disputes. Their answer to the uncertainties of a technological society was not to reject technology but to return to fundamentalist religion and traditional beliefs. It was in this context that the NSF precollege science curriculum became the focus of prolonged and bitter attack.

Scientists tend to associate the questioning of scientific rationality with alienation, crackpot ignorance, or pathology. They label those who persist in their criticism of scientific explanations as "irrational," "marginal," "out of touch" with modern society. The battles over science textbooks in the 1970s, however, cannot be easily dismissed as social pathology, for the challengers are people far removed from the deprived and marginal subcultures long associated with pentecostal sects; they are sober and solid, often technically trained, middle-class citizens, employed in white collar jobs. Their ideas are squarely in line with Horatio Alger myths of individualism and self-determination, for they are seeking order, certainty, and some measure of independent control over their lives. Many of the textbook watchers identify themselves as scientists, arguing that alternative Biblical theories can be scientifically validated. If the conservative groups who participate in these movements feel a sense of deprivation, it is their sense of decreased power, due to loss of local control to technical bureaucratic institutions which, they feel, do not represent their interests. If they are marginal, their marginality is more philosophical than social, for they are engaged in an ideological battle to preserve their values against "the forces of irreligion, agnosticism, and secularism."

Yet the textbook disputes are less a manifestation of antiscience attitudes than a reflection of a much broader concern in contemporary American society with the social, political, and cultural implications of science. Criticism of science textbooks expresses the growing tensions in the relationship between science and society as public confidence in the unmitigated utility of science wanes. Similar to many disputes over such diverse scientific and technical developments as fluoridation, nuclear power, or biomedical and genetic research, textbook controversies expose concerns about the increasing role of technical expertise, the dominance of professional bureaucracies, and the neglect of local interests, and they suggest how such concerns can be translated into political and administrative decisions.

This study explores the complex spectrum of motives and perceptions underlying contemporary criticism of science as expressed in the arena of public education. Part I briefly portrays the historical context of present-day disputes, discussing those aspects of the early reaction to Darwinian evolution and its introduction in the schools that have recurred in contemporary conflicts. To provide more background on the textbook conflicts, Part I also describes the science curriculum reform movement, while Part II discusses contemporary textbook watchers, analyzing their social and political base, and Part III describes their activities in several disputes. Finally, Part IV analyzes the disputes in terms of the social and political concerns underlying the collision of values over science education, and the relationship between science and personal beliefs.

I THE CONTEXT

[Science] wrote an end to the ancient animist covenant between man and nature, leaving nothing in place of that previous bond but an anxious quest in a frozen universe of solitude. With nothing to recommend it but a certain puritan arrogance, how could such an idea win acceptance? It did not; it still has not. It has however commanded recognition; but that is because, solely because, of its prodigious power of performance. —Jacques Monod, *Chance and Necessity*

2 A Science or a World View? Historical Notes on the Teaching of Evolution

The metaphysical assumptions and moral implications inherent in aspects of evolution theory have been a source of innumerable battles for over a hundred years. Pre-Darwinian biologists based their science on theological assumptions. Science was rooted in religion; its purpose was to prove the existence of God, using as evidence the design and purpose in nature. Darwin introduced an explanation of biological change that excluded the necessity of supernatural intervention and incorporated elements of chance and indeterminacy. Thus, Darwin's *Origin of Species* was viewed as a revolutionary document in 1859, although its primary contribution was to organize and synthesize a set of ideas that had pervaded the scientific literature for more than fifty years.[1] A brief review of the reactions to evolution theory in the nineteenth century, and later to its introduction in the American school system, suggests the persistence of certain concerns about its underlying assumptions and social implications.

Evolution and the Nineteenth-Century Soul

By 1800, theological explanations of human origins such as those of Archbishop James Ussher, the seventeenth-century prelate who established the year of origin as 4004 B.C., were already being seriously questioned as studies in biology and geology generated a variety of evolutionary hypotheses. For example, as early as 1796, James Hutton had attributed geological features to physical causes, introducing the principle that uniform geological agents operated during all periods of history to provoke continuous change.[2] By the 1830s, Charles Lyell had elaborated this uniformitarian hypothesis and laid the groundwork for modern geology by describing the development of the earth over many ages.[3] Other evolutionary theories appeared; the concept of catastrophism, associated mainly with Georges Cuvier, suggested that change occurred by sudden cataclysms that destroyed life and gave rise to new waves of increasingly complex living forms. Later, Lamarck related the tendency toward increased complexity to the influence of environment, which, he claimed, caused adaptive changes that were transmitted to subsequent generations. Others, such as Karl Nägeli, postulated that inner-directed forces guided evolution.[4] Meanwhile, a body of rich geological data accumulated.

Darwin saw evolution as the interplay of three principles: the occurrence of random variation; the mechanism of heredity, which transmits similar organic forms; and the struggle for existence, which determines which variations survive to be inherited. He described natural selection as a plausible causal mechanism for evolution in an environment that directly determines the survival of better-fitted variants (those able to reproduce themselves). Later, neo-Darwinians, benefiting from better understanding of genetics and the statistical analysis of gene pools, would develop evolution theory more in terms of the statistical transformation of populations than of changes in individuals, but in the nineteenth century Darwin's work provided coherence and organization to a mass of observations and permitted scientists to develop logical postulates concerning the origin of life billions of years ago.

Despite its rich empirical base and the prior history of evolutionary thought, *The Origin of Species* was profoundly disturbing, for concepts of unlimited change and purposelessness, the "twin philosophical implications of Darwinism" distressed the nineteenth-century soul. While evolutionary concepts had been implicit in previous scientific advances, "Darwin brought them into the open when he applied them to the hitherto sacrosanct topic of the origin and destiny of man."[5]

The most active nineteenth-century resistance to the new science came from scientists themselves. The traditional scientific view was locked into a theoretical system based on the assumption that science must justify "the divine presence" and that scientific theories must rest on "a primordial creative power." The historian Charles Gillispie talks of the nineteenth-century debate as "one of religion in science rather than religion versus science."[6] Asa Gray, for example, saw the need of theories to prove an overarching design or ultimate purpose. Louis Agassiz described the obligation of the scientist: "Our task is . . . complete as soon as we have proven His existence."[7]

Agassiz accepted the concept of evolutionary change, but assumed that the members of every species shared a common essence and that the essential character of any species was immutable. Natural selection violated this assumption of "fixity of species," as it violated earlier theories of uniformitarianism and catastrophism. Moreover, the idea that blind physical forces could generate change invalidated the assumption of progress toward an ideal state of perfection.

Perceiving science as an inductive process in which the collecting of facts was the only possible basis for constructing a theory, Darwin's critics also attacked him for violating scientific methodology. They argued that evolution was "mere hypothesis; unsubstantiated speculation." There were no concrete facts to demonstrate the existence of random variation, nor were there known transitional links between species. Adam Sedgwick, professor of geology at Cambridge, detested the theory "because it has deserted the inductive track, the only track that leads to physical truth; because it utterly repudiates final causes and thereby indicates a demoralized understanding on the part of its advocates."[8] And Sir Richard Owen, a paleontologist from the British Museum, expressed his disappointment about the inadequacy of observations: "We are called upon to accept a hypothesis on the plea of want of knowledge."[9] Darwin replied to such criticism by defining his sense of scientific method. "I am actually weary of telling that I do not pretend to adduce direct evidence of one species changing into another, but that I believe that this view is in the main correct because so many phenomena can thus be grouped together and explained."[10]

Beyond its impact on traditional science, Darwinism was devastating to conventional theology. Religious traditionalists accused Darwin of "limiting God's glory in creation," of "attempting to dethrone God," of "implying that Christians for nearly 2,000 years have been duped by a 'monstrous lie.'"[11] Evolution theory violated traditional theological assumptions, and, above all, the assumed distinction between man and the animal world. Yet Darwin himself did not reject the concept of a creator, writing that "There is grandeur in this view of life, with its several powers, having been originally breathed by the Creator into a few forms or into one; and that . . . from so simple a beginning endless forms most beautiful and most wonderful have been and are being evolved."[12]

By the turn of the century, both scientists and theologians were increasingly inclined to accept Darwinism as revealing God's purpose, and some elaborate theories attempted to reconcile geology and Genesis. P. H. Gosse, the religious naturalist, developed the omphalos theory that denied there had been a gradual modification of the earth. "When the catastrophic act of creation took place, the world presented instantly the structural appearance of a planet in which life had long existed."[13] Efforts at reconciliation included suggestions that the six days of creation were indeed eras, or that God created fossils simply to test man's faith.[14] Years later, Catholic theologians would reconcile science and religion as "two different approaches to reality, distinct in their methods of thought."[15] Both were concerned with the search for an orderly harmonious universe, but neither excluded the other. By a sort of truce, it was assumed that religion provided a vision of a world beyond nature, while science was grounded in reality. Efforts to demonstrate the compatibility of religious conviction with modern science, to coordinate "the reality of divine revelation in the Cosmos and in Scripture," to prove that "Christ and not evolutionary process is the only adequate index to cosmic activity and purpose"[16] continues to this day.

By the end of the nineteenth century, the scientific and religious implications of Darwinism were fully matched if not exceeded by its philosophical and social influence. Late nineteenth-century entrepreneurs such as John D. Rockefeller and Andrew Carnegie found natural selection a comfortable sanction for laissez-faire economics. "The growth of large business is merely the survival of the fittest." Industrial competition was not an evil tendency, but "merely the working out of a law of nature and a law of God."[17] Indeed, postbellum America could be described as a caricature of the struggle for existence and survival of the fittest.

Evolution had an extraordinary appeal as a vision of reality. Herbert Spencer, the "cosmic evolutionist," soon applied the concept to society.

The advance from the simple to the complex, through a process of successive differentiations, is seen alike in the earlier changes of the

Universe to which we can reason our way back . . . it is seen in the unfolding of every single organism of its surface, and in the multiplication of kinds of organisms; it is seen in the evolution of Humanity, whether contemplated in the civilized individual, or in the aggregate of races; it is seen in the evolution of Society in respects alike of its political, its religious, and its economic organization. . . .[18]

Later, Julian Huxley described the "evolutionary vision" as a naturalistic religion,[19] and C. H. Waddington, the noted British biologist, saw evolution as "a secure basis for ethics."[20]

In the United States, Victorian culture absorbed the theory of evolution.[21] In 1895, the National Education Association recommended a zoology course that was evolutionary in organization, and biology textbooks prior to 1920 began to introduce evolution theory to secondary schools and college students, often presenting it with extraordinary and perhaps self-conscious assurance. "We do not know of any competent naturalist who has any hesitation in accepting the general doctrine."[22] "There is no rival hypothesis to evolution, except the out-worn and completely refuted one of special creation, now retained only by the ignorant, dogmatic, and the prejudiced."[23]

Antievolution in the 1920s

The reconciliation between evolution and religion was not to last in the United States; the conflict became a public issue in the 1920s—provoked by fundamentalists in the South. Fundamentalism was an American movement that developed around the turn of the century, and expanded with a publication in 1909 of a series of pamphlets entitled *Fundamentals*.[24] These pamphlets attacked "modernism" and, in particular, evolution theory; the idea that evolution involved discrete accidental changes determined by the circumstances of the moment shattered fundamentalist faith in planned and purposeful change. They attacked the theory with zeal and enterprise, and worried about its implications for Christian behavior.

By the 1920s, the issue of evolution divided Protestant churches; fundamentalists denied the validity of evolution and modernists sought to reconcile their faith with science. Control over educational institutions was a major arena for their battles. The major fundamentalist reaction against evolution took place in "backwoods" areas, among people inclined to be irritated by the liberal and disdainful attitudes of the industrial North. Intellectuals of the time viewed it as a populist, anti-intellectual movement—"a revelation of the backwardness and intolerance of large elements of our population."[25] Indeed, among Northerners who had reconciled religion and evolution, the old assumption about the incompatibility of science and religion seemed almost absurd by the time the Scopes trial brought the issue to national prominence. In fact, the trial was not intended to raise this issue at all. Rather, it was provoked by the American Civil Liberties Union in order to show that Tennessee's antievolution legislation violated the First Amendment. The constitutional question, however, was buried as William Jennings Bryan and Clarence Darrow clashed over questions of religion and morality, and conveyed the resentment between northern liberals and southern conservatives.

Arthur Garfield Hays, who worked with Darrow for the defense, could hardly believe that "religious views of the middle ages" could recur "in spite of railroads, steamboats, the World War, the telephone, the radio, the airplane, all the great mechanistic discoveries. . . . "[26] Yet fundamentalists had considerable political influence in the twenties. Between 1921 and 1929, antievolution bills were introduced into thirty-seven state legislatures, and there were increasingly successful efforts to qualify statements about evolution in those textbooks that included discussion of the theory, or to exclude such textbooks from adoption in the schools.[27]

One contemporary writer described the prevailing emotional hostility to science as a "cancer of ignorance," a repudiation of the authority and integrity of scientists.[28] But hostility was due largely to the association of evolutionary theory with disturbing social problems of the day. Evolutionary ideology, claimed Bryan, went beyond simple scientific questions and bore on moral values. The force of this argument was great in the 1920s, a comparatively lawless period; popular discussion suggested that the country was "going to ruin." Harry Emerson Fosdick, a liberal Baptist, speaking for "a large number of Christian people" disagreed with many of Bryan's ideas, but fully concurred that "Everyone closely associated with the students of our colleges and universities knows. Many of them are sadly confused, mentally in chaos, and, so far as any guiding principles of religious faith are concerned, are often without chart, compass or anchor."[29] Social problems were variously attributed to the weakening of loyalty to the church, to postwar letdown, and to prosperity. But blame for "immorality" was also laid on a materialism that was fed by science and especially by the teaching of evolution.

The Scopes trial reflected parental demands to control the education and the values of their own children. "What right," Bryan asks, "have the evolutionists—a relatively small percentage of the population—to teach at public expense a so-called scientific interpretation of the Bible, when orthodox Christians are not permitted to teach an orthodox interpretation of the Bible?"[30] Scientists, however, had quite a different view about populist control of science curricula. "What is to be taught as science would be determined not by a consensus of the best scientific opinion, but by the votes of shopgirls and farmhands, ignorant alike of science and of the foundation principles of our civil society."[31] While the people of Dayton, Tennessee, defined the issue in terms of their social and moral concerns, the proponents of teaching evolution focused on the rational and logical basis of the theory. Clarence Darrow, for example, tried to force Bryan to a literal defense of the Bible as a scientific and logical document in order to "show up fundamentalism . . . to prevent bigots and ignoramuses from controlling the educational system."[32]

Although Scopes lost his case, the trial ended for a while the public efforts of fundamentalists to ban the teaching of evolution. The textbook crusades that were so influential in the 1920s faded from public view during the Depression, as economic problems and support for prohibition occupied the time of fundamentalist leaders. Antievolution sentiment persisted, but mostly among millennial sects such as Seventh Day Adventists and Jehovah's Witnesses. Their tracts continued to denounce evolution theory as both incorrect and responsible for "the progressive worsening of crime, delinquency, immorality and even war . . . morals are broken down and for multitudes, faith in God has been shattered . . . evolution has paved the way for an increase in agnosticism and atheism as well as opening the door for communism."[33] During the 1940s and 1950s, fundamentalists were preoccupied with maintaining their own subculture, setting up Bible camps, colleges, seminaries, newspapers, and radio stations. To the extent that they attacked the public schools, they focused more on prayers and sex education that on evolution.[34]

The Legacy of the Scopes Trial

The relative quiescence of fundamentalists concerning the teaching of evolution in part reflected the neglect of the subject in biology textbooks after the Scopes trial. A scholarly survey of the content of biology texts up to 1960 found the influence of antievolutionist sentiment to be persistent, if undramatic, and showed that the teaching of evolution actually declined after 1925.[35] Textbooks published throughout the late 1920s ignored evolutionary biology, and new editions of older volumes deleted the word evolution and the name Darwin from their indexes. Some even added religious material. By the late 1930s some publishers were tentatively introducing evolution, but most, discouraged about market prospects and anxious to avoid controversy, avoided the topic, focusing largely on morphology and taxonomy.[36] In 1942, a nationwide survey of secondary school teachers indicated that fewer than 50 percent of high school biology teachers were teaching anything about organic evolution in their science courses. Some avoided the subject either because of potential community opposition or their own personal beliefs. At the time, a biologist claimed that "biology is still pursued by long shadows from the middle ages, shadows screening from our people what our science has learned of human origins . . . a science sabotaged because its central and binding principle displaces a hallowed myth."[37] Fifteen years later, biology courses still neglected paleontology and evolution. A major biology textbook devoted only one page to evolution theory, which it called "racial development," and avoided discussion of origins.[38] As late as 1959, Herman J. Muller could write that public school biology teaching was dominated by "antiquated religious traditions."[39] One hundred years after the theory of evolution by natural selection was firmly established, it was still not an integral part of the public school curriculum.

Meanwhile, laws forbidding the teaching of evolution in public schools remained on the books in several southern states. In January 1961, a bill to repeal Tennessee's monkey laws, still in force thirty years after the Scopes trial, met prompt and passionate rejection by people who argued that evolution theory "drives God out of the universe" and "leads to communism." "Any persons or any groups who assist in any way to undermine faith in the teaching of the Bible are working in harmony with Communism."[40] In 1967, Gary Scott, a high school teacher in Jacksboro, Tennessee, was dismissed from his job for violating the state statute. He appealed with legal assistance from the National Science Teachers Association. By this time reapportionment had given greater representation to urban areas in Tennessee, and the law was finally repealed. The sponsor of the repeal measure, D. O. Smith from Memphis, brought to the state legislature a caged monkey with the sign reading "Hello, Daddy," but his humor and the subsequent repeal of the legislation did not mean that the values of Tennessee citizens really changed. This was to become evident when the legislature met again on the issue in 1973.

In Little Rock, Arkansas, Governor Faubus defended antievolution legislation throughout the sixties: it was "the will of the people." But Susanne Epperson, a high school teacher, challenged the legislation and won a favorable judgment in a local court. "The truth or the fallacy of arguments on each side of the evolution debate does not either contribute to or diminish the constitutional right of teachers and scientists to advance theories and to discuss them."[41] The state supreme court, however, upheld the constitutionality of the law forbidding the teaching of evolution, and it was not until 1968 that the U. S. Supreme Court ruled the Arkansas antievolutionary law unconstitutional. Following this precedent, the last of these laws (in Mississippi) was soon off the books.

Even as the courts challenged the old legislation forbidding the teaching of evolution, textbook watchers were gathering momentum for a renewed attack on the evolutionary assumptions of the new biology and social science courses being introduced in their public schools.

Notes

1. Bentley Glass et al., eds., *Forerunners of Darwin 1945–1859* (Baltimore: Johns Hopkins Press, 1959).

2. James Hutton, *Theory of the Earth* (London: 1796).

3. Charles Lyell, *Principles of Geology* (London: John Murray, 1835).

4. For selections from Darwin's forerunners and a history of the response to Darwinism, see Philip Appleman, ed., *Darwin* (New York: W. W. Norton, 1970).

5. Langdon Gilkey, "Evolution and the Doctrine of Creation," in *Science and Religion*, ed. Ian Babour (New York: Harper & Row, 1968), p. 164.

6. Charles Gillispie, *Genesis and Geology* (Cambridge, Mass: Harvard University Press, 1951). For discussion of this point, see also Ernst Mayr, "The Nature of the Darwinian Revolution," *Science*, 2 June 1972, pp. 981–989.

7. Louis Agassiz, *Essays on Classification* (Boston: Little, Brown, 1957), p. 132.

8. Adam Sedgwick, "Objections to Mr. Darwin's Theory of the Origin of Species," *The Spectator*, 24 March 1860, p. 286.

9. Review of *The Origin of Species* in *Edinburgh Review* CXI (1860), reprinted in Appleman, *Darwin*, pp. 295–298.

10. Francis Darwin and A. C. Seward, eds., *More Letters of Charles Darwin* (London: S. Murray, 1968), vol.I, p. 184.

11. Andrew Dickson White, *A History of the Warfare of Science with Theology in Christendom* (New York: 1896).

12. Charles Darwin, *The Origin of Species* II, (New York: P. F. Collier, 1902), p. 316.

13. Edmund Gosse, *Father and Son* (London: Heine, 1907), p. 100.

14. For discussion of the reconciliation between science and religion, see Ian Barbour, *Science and Religion* (New York: Harper & Row, 1968), pt.II; Henry Drummond, "The Contribution of Science to Christianity," in *Henry Drummond, an Anthology*, ed. J. W. Kennedy (New York: Harper & Row, 1973); John Dillenberger, *Protestant Thought and Natural Science* (New York: Doubleday, 1960).

15. See, for example, discussion in Paul Chauchard, *Science and Religion* (New York: Hawthorn Books, 1962), p. 148.

16. Carl F. H. Henry, "Theology and Evolution" in *Evolution and Christian Thought Today*, ed. Russell Mixter (Grand Rapids, Michigan: Eardmans, 1959), p. 202.

17. Cited in Richard Hofstadter, *Social Darwinism in American Thought* (Philadelphia: University of Pennsylvania Press, 1955), ch.2.

18. Herbert Spencer, "Progress: Its Law and Causes" in *Essays: Scientific, Political and Speculative* (New York: Appleton, 1915), p. 35.

19. Julian Huxley, *Evolution in Action* (New York: Harper & Row, 1953), and *Religion without Revelation* (New York: Harper, 1927).

20. C. H. Waddington, *The Scientific Attitude* (London: Penguin Books, 1941), and *The Ethical Animal* (London: G. Allen and Unwin, 1960).

21. There was, however, a continued current of opposition to science expressed, for example, in the antivivisection movement. See Richard D. French, *Anti-Vivesection and Medical Science* (Princeton: Princeton University Press, 1975).

22. J. Arthur Thomson, *Concerning Evolution* (New Haven: Yale University Press, 1925), p. 53.

23. H. H. Newman, *Outlines of General Zoology* (New York: Macmillan, 1924), p. 407.

24. The movement was rooted in the ethos of Methodist revivalism that developed in the early 1800s, but it later crystallized around the issue of Darwinism.

25. S. J. Holmes, "Proposed Laws Against the Teaching of Evolution," *Bulletin of the American Association of University Professors* 13, No. 8 (December 1927), pp. 549–554.

26. Arthur Garfield Hays, "The Scopes Trial," in *Evolution and Religion*, ed. Gail Kennedy (New York: D. C. Heath, 1957), p. 36.

27. Antievolution laws were passed in Mississippi (1926), Arkansas (1928), and Texas (1929).

28. Chester H Rowell, "The Cancer of Ignorance," *The Survey*, 1 November 1925.

29. Harry Emerson Fosdick, "A Reply to Mr. Bryan in the Name of Religion," in Kennedy, *Evolution and Religion*, p. 33.

30. Bryan's ideas on evolution theory were influenced by George McCready Price.

31. Holmes, "Proposed Laws Against the Teaching of Evolution," p. 554.

32. Frederick Lewis Allen, *Only Yesterday* (New York: Blue Ribbon Books, 1931), p. 205.

33. Jehovah's Witness, *Did Man Get Here by Evolution or Creation?*, Watchtower Bible and Tract Society, Brooklyn, 1967, pp. 168–170.

34. The Christian Crusade, run by Preacher Billy James Hargis, published a booklet called "Is the Little Red Schoolhouse the Proper Place to Teach Raw Sex?" This caused the demise of many sex education programs in the United States. Other fundamentalist demagogues such as Gerald L. K. Smith associated fundamentalism with nationalism.

35. Judith Grabiner and Peter Miller, "Effects of the Scopes Trial," *Science*, 6 September 1974, pp. 832 ff.

36. Cornelius Troost, "Evolution in Biological Education Prior to 1960," *Science Education* 51 (1967): 300–301.

37. Oscar Riddle, "Preliminary Impressions and Facts from a Questionnaire on Secondary School Biology," *The American Biology Teacher* 3 (February 1941): 151–159. This survey, done by the Union of American Biological Societies, was sent to 15,000 teachers, of whom 3,186 responded.

38. Truman J. Moon et al., *Modern Biology* (New York: Henry Holt, 1956), p. 665.

39. Hermann J. Muller, "One Hundred Years without Darwinism are Enough," *The Humanist* XIX (1959): 139.

40. W. Dykeman and J. Stokeley, "Scopes and Evolution—the Jury is Still Out," *New York Times Magazine*, 12 March 1971, pp. 72–76.

41. Quoted in an editorial, *Science Teacher* 33 (September 1966): 17.

3 The Science Curriculum Reform Movement

As Sputnik was launched in 1957, so was the science curriculum reform movement, intended to build up scientific and technical manpower by bringing modern scientific knowledge to the nation's public schools. Science curriculum reform was an effort to enlist the public education system in the resolution of the problems of the cold war. It was based on the assumption that the public school curriculum should introduce students to the key concepts and methods of the academic scientific disciplines.

Nearly every social and political cause looks to the school system as a means of effecting social reform. The demand for skilled labor in a growing industrial economy had prompted the movement for manual training in the schools. Concerns with urban migration and the bleak future of agriculture generated the Nature Study Movement. The progressive education movement after World War I reflected Wilsonian visions of a new and better world; schools became "an engine for social benefit," a means to apply the promise of American ideals to industrial civilization.[1]

The new science courses grew out of disillusion with the progressive movement, although they shared many of its ideals. "Our technological supremacy has been called into question," declared Admiral Hyman Rickover. "Parents are no longer satisfied with life adjustment schools. Parental objectives no longer coincide with those professed by professional educationists."[2] American schools, it was discovered, lagged behind their European counterparts, and textbooks hardly reflected the rapid development of modern scientific knowledge. The fragmented and competitive textbook industry, trying to reach a national market, seemed unwilling to run the risk of presenting new science materials for which there was little existing demand. Concerned scientists turned to federal agencies for what was, after all, a problem with national implications.

The Federal Government's Role

The public school system in the United States is anarchic: about 20,000 school districts operate as autonomous domains serving a diverse population from rural, suburban, and inner city backgrounds. Anarchy permits flexibility in educational experimentation, as long as it is not costly. However, the diversity of values and differences in priorities in this complex school system limits costly or controversial innovation.

It is customary to deal with diversity by avoiding conflict. For school teachers this has meant a considerable reluctance to introduce subjects such as evolution theory in classrooms where religious objections are likely to appear.[3] For the highly competitive textbook industry, conflict avoidance means continual compromise. The industry must produce books that are provocative, but not so different as to be controversial; textbooks must stand out in some way in order to compete effectively, yet be sufficiently standard to attract the largest possible market. Most publishers, operating on a low profit margin, shun experiments and avoid controversy.[4] Major textbook reforms are, therefore, seldom initiated by the industry, and in subject areas of high relevance to national goals, federal agencies felt it appropriate to intervene. This was the situation when competition with Soviet technology brought awareness of the need for technical manpower.[5]

In 1957, Jerrold Zacharias and a group of physicists from Cambridge, Massachusetts, formed the Physical Science Study Committee (PSSC). This group proposed to develop films to teach physics to high school students, and they received a grant of $303,000 from the National Science Foundation to develop the project. PSSC was the first course in the NSF's science curriculum development program, which soon grew into an important activity within the Foundation, and a major enterprise in the history of education.[6] The physics program was followed by mathematics, chemistry, biology, and finally by social sciences; by 1975, NSF had funded fifty-three projects at a total cost of $101,207,000. (See Appendix 1.)

Following the model developed by the PSSC, research scientists worked with educational experts to prepare course materials that reflected the knowledge and methods of modern science.[7] The new curriculum thus tried to remedy the authoritarian textbook presentation of science. Students had long been fed "objective facts" and "proven laws of nature" through textbooks based on the assumption that science was simply an inductive process. Scientists themselves had rejected this approach; the role of theory was less to explain than to predict. Students, claimed the curriculum reformers, must therefore be taught to use theory as an instrument of prediction and to participate in the process of scientific discovery. The new curriculum focused on the methods of science and introduced students to the research process. Pedagogical techniques followed from the idea that the student must be an active "scientific investigator" rather than a passive recipient of materials provided by teachers. Moreover, the materials were to be "teacher-proof," conveying accepted concepts of science regardless of local opinions or local circumstances. The science courses thus not only updated materials but tried to change the methods of education, focusing less on the learning of facts than on the methods of inquiry and the exercise of individual judgment.

At first the NSF approached the problem of improving science education by supporting projects to develop new textbooks, films, and audiovisual materials as proposed by various educational centers and university departments throughout the country. It very quickly became apparent that the development of courses or books was not sufficient to assure their utilization. Due to the costly nature of multimedia materials and equipment, an initial federal subsidy was needed to implement their use. Teachers had to be trained to understand the new curriculum and to use the new pedagogical techniques.

NSF approached programs for implementation with some reluctance. While concerned with disseminating the material it had sponsored, Foundation officers were also aware of the political sensitivity of local school districts to any implementation program that might appear to be "federal intervention." They were constantly under congressional pressure to avoid such "interference."[8]

"You do not interfere unless they ask you to interfere." (Senator Magnuson, 1958)

"I don't want these things rammed down the throat of educators."
(Senator Allott, 1964)

"You are not suggesting the philosophy . . . that is being taught?
This is a local option of the school district." (Congressman Talcott,
1967)

Federal funds were finally appropriated for three types of implementation activities: leadership projects to train specialists who
would influence curriculum decisions and guide local implementation efforts; teacher training projects;[9] and projects to help local
schools, in cooperation with colleges and universities, put NSF materials and techniques into practice. (See Table 1.)

Until the late 1960s the thrust of the new NSF-supported curriculum was to encourage scientific careers. An important early exception was the Biological Sciences Curriculum Study (BSCS) which was
directed to a much wider student body and essentially taught "biology for the citizen." Later in the sixties, responding to a general social concern with "science for citizenship," the Foundation began to
shift its emphasis in other courses as well. The declining need for
scientists and engineers had coincided with an apparent disillusion
with science and technology and their association with social and environmental problems. Better public understanding of science
might, it was hoped, create more favorable attitudes. Again, the
public schools were the vehicles for reform, and curriculum was now
to be directed not only to future scientists but to future citizens in a
society in which science was a crucial economic, social, and political
influence. This change in perspective, however, also required new
pedagogical techniques to make science more palatable to nonspe-
cialized students. Science teaching would have to be oriented
around practical problems rather than traditional disciplines, and
this implied increased attention to the social and behavioral sciences.

NSF had always approached the support of social science research
and education with caution. The physical scientists who had formed
the Foundation during the late 1940s and shaped its development
had been strongly opposed to the inclusion of social science among
the Foundation's activities. They argued that the quality of work in
the field, and its political connotations of socialism and public manipulation would undermine public confidence in other scientific

Table 1. NSF Support of Four Precollege Projects for Education in the Sciences in Thousands of Dollars (Fiscal Years 1958-1974)

Program Category	1958	1960	1962	1964	1966	1968	1970	1972	1974
Course Content Improvement Programs	835	6,299	8,990	13,976	15,564	19,352	9,840	10,698	8,261
Secondary School Student Programs	389	4,458	2,898	3,198	1,973	2,067	1,931	1,938	1,375
Institutes—Precollege Teachers	12,212	33,775	40,876	43,247	40,531	38,328	39,866	23,373	17,143
Cooperative College-School Science Programs	—	—	521	721	1,957	3,387	4,654	4,355	—

Source: National Science Foundation Science Curriculum Review Team, *Pre-College Science Curriculum Activities of the NSF, II*, May 1975, p. 43.

activities.[10]

The Foundation eventually funded social science research under the vague rubric, "And Other Sciences," but it remained extremely careful to support only "ultra-safe lines of inquiry."[11] Only in 1968, when Congress endorsed the Division of Social Sciences through an amendment to the NSF Act, were the social sciences afforded status equal to other categories of science. Congress was especially cautious with respect to curriculum development. By this time, however, the NSF officers were complacent about the social sciences: in the budget hearings for fiscal year 1966, Dr. Riecken asserted, "The kind of social science materials being produced under our grants are not, I think, likely to be as much trouble as the biologists have had over evolution."[12]

Complacency was understandable. Until the late 1960s science education was relatively immune to attack—with the important exception of persistent efforts to keep evolution out of textbooks. And during the years following World War II, when textbook watchers were most active, even this issue received relatively little attention compared, for example, to questions of patriotism or sex education. The development of the science curriculum, a highly professionalized activity, was hardly a problem to be thought of in political terms. The earliest NSF courses in the physical sciences, developed for a limited group of potential scientists, raised few controversial issues. If there were profound objections, faith in science and its association with material progress overwhelmed them.

Scholars had enthusiastically acclaimed many of the new course materials, and once federal initiatives stimulated a market for these materials, private publishers followed, incorporating many of their features in their own publications. By 1970, about one fourth of all high school students in the United States were taking science courses based on NSF materials.[13] The program thus brought about widespread pedagogical change, with implications for teachers, students, parents, publishers, testing agencies, local school boards, and other participants in the educational system. Yet the Foundation continued to treat curriculum development contracts much as it treated research grants, relying on the professional peer review system to control the quality and accuracy of the work performed under its sponsorship. As in the case of funded research, this meant minimum interference with the course development process.[14] The Foundation thus supported curriculum development, but refused to endorse or claim responsibility for the educational value of the materials produced. Its role was to make materials available but not to mandate their use. This was a deliberate policy intended to avoid accusations of federal intervention in local educational policy.

Inevitably, however, dissemination of federally funded materials increased the tension between the federal government and local school districts jealously guarding their autonomy; and between professional educators concerned with disseminating knowledge and parents preoccupied with prepetuating family values. As the science curriculum, reflecting contemporary research, dealt with controversial aspects of evolution theory and its importance for understanding human development and behavior, these tensions increased. By improving the public school curriculum, the NSF, despite its denial of responsibility, found itself involved not in an isolated neutral research endeavor but in a major social intervention filled with political implications.

Two New Courses

The Biological Sciences Curriculum Study

Soon after the NSF funded the physics program, the American Institute of Biological Sciences (AIBS) began to discuss the problem of public education in biology, which appeared to be at least twenty years behind current research in the discipline. Biologists were particularly concerned about the neglect of evolution theory. Population genetics, entomology, and experimental field work in paleontology and geology had brought both refinements to and increasing support for evolution theory. It was confirmed and reconfirmed as a useful scientific hypothesis. Yet, during the Darwin centennial celebration of 1959, a group of selected teachers at an NSF summer institute agreed that teaching evolution was still hampered by deficient texts, inadequate knowledge among schoolteachers, and their continued fear of local religious opposition.

Biologists were appalled to realize that the sophisticated research in the discipline was so poorly taught. Biology is, after all, a key course in general science education. Over 90 percent of American high schools offer a biology course, and an estimated 80 percent of all high school students enroll in the subject; for most, it is their only exposure to science. After the centennial, a group of distinguished scientists from the AIBS formed the Biological Sciences Curriculum Study (BSCS) at the University of Colorado to develop a modern approach to the teaching of biology. NSF provided the center with $7 million.

Unlike physics and chemistry, biology had to be presented at a level where it would be studied by a majority of students.

A sound biological understanding is the inalienable right of every child who, when adult, will need to cope with individual problems of health and nutrition; with family problems of sex, and reproduction and parenthood; and with citizens' problems of wise management of natural resources and biological hazards of nuclear agents. . . . [15]

BSCS decided to abandon the earlier taxonomic approach requiring extensive memorization, and instead to focus on methods and laboratory research. "Inquiry-oriented" instruction replaced the "pickled frog approach" involving memorization of facts. "The emphasis . . . is placed on the nature of science, on the men who have worked as scientists, and on scientific inquiry. The point is to teach science not as a body of classified knowledge, but as an approach to problem solving."[16]

Finally, the BSCS determined that their work would not avoid controversial subjects such as organic evolution and that they would base the new curriculum on current themes of biology: the change of living things through time, the diversity of types and unity of pattern in living things, the genetic continuity of life, the complementarity of organisms and environment, and the biological roots of behavior.[17]

BSCS set up curriculum study groups involving professional biologists and educators. These had no commercial dependence on the nonacademic marketplace, for BSCS felt strongly that biology books had in the past suffered from the influence of political interest groups and apprehensive public officials.

In 1963, after nearly five years of preparation and testing, BSCS marketed three introductory textbooks for high school classes. Each volume had a slightly different emphasis, reflecting trends in biological research; one stressed cellular biology, another ecology, another molecular analysis.[18] However, the material overlapped by about 70 percent, and each book was based on evolutionary assumptions as "the warp and woof of modern biology."[19]

In 1961, as a part of its testing program, BSCS had tried out one of the books in Dade County, Florida, and had run into a skirmish when school authorities asked that diagrams of the reproductive system be removed. BSCS refused and the school officials themselves blackened out offending material. BSCS scholars were not particularly surprised at this reaction, but they did not anticipate a religious opposition. It was generally assumed that, while an occasional extremist might emerge, sympathy for scientific accuracy would prevail.

When the books were finally marketed, there were a number of isolated incidents of hostility. School supervisors in several southern states refused to purchase the books. The state board of education of New Mexico insisted that the inside front covers of all BSCS books be stamped with a note emphasizing that evolution was a theory, not a fact, and that this was the official position of the state board. Then a major dispute took place in Texas, a state with a long tradition of conservatism in the selection of textbooks. When BSCS books were submitted for adoption to the state textbook screening committee, the Reverend Reuel Lemmons, of the Church of Christ, a fundamentalist sect claiming 600,000 members in Texas, appealed to the board of education and to Governor John Connally. He called the books "pure evolution from cover to cover, completely materialistic and completely atheistic," and asked his constituents to petition against their adoption.

The campaign against BSCS in Texas continued during the summer and fall of 1964. One critic suggested in a letter to Connally that "the assassination of our late president and the attempt on your life were the vicious attacks of a Godless individual. It is the purpose of this letter to call your attention to a situation in our state which is designed to promote such Godlessness in our public schools through the atheistic teaching of evolution theory."[20] That the assassination took place from the state textbook depository was not ignored by textbook opponents.

The protest reached its peak in October when the state textbook selection committee held a public hearing. Lemmons and his supporters, including two college professors, sought to have the textbooks banned as a violation of the First Amendment. The committee finally approved all three BSCS books, but only after several changes in the books that softened their evolutionary emphasis. These, for example, specified that evolution theory was a theory, not a fact, and that it had been "modified," not "strengthened" by recent research.

Subsequently, nearly 50 percent of American high schools used BSCS materials, which later included films, pamphlets, and other instructional programs. During most of the 1960s, the major problem facing BSCS was less a matter of social protest than the inertia of high school teachers, who often failed to understand the materials and the methods of science sufficiently to convey the character and use of evolution theory in biology.

Man: A Course of Study
Social studies courses for upper elementary school children often consist of descriptive presentations of American history directed toward creating "responsible citizens" with allegiance to the values of democracy. Few students are exposed to the study of human behavior until their college years. That this situation was perceived as a problem in the 1960s reflected the contemporary confidence (also evident in the Great Society programs) that social science knowledge could be useful in coping with major social problems.

In the early 1960s, the American Council of Learned Societies began to grapple with reform of the social studies curriculum. To meet potential objections, the Council first had to reconcile the "new" social sciences with the well-entrenched concepts and goals of the traditional social studies curriculum. It concluded that social science did not conflict with traditional goals, but rather would enhance them by developing the student's analytic perspectives and respect for "objective knowledge" about human behavior. The controversial issues that were bound to arise could be dealt with in a spirit of open scientific inquiry free from prejudice and ignorance.

Reflecting this belief, in 1963 a group of scholars organized by Jerome Bruner met in Cambridge, Massachusetts, with the Educational Services Institute (later called the Education Development Center (EDC)) to discuss the possibility of developing a new humanities and social science curriculum. The result of this meeting was an exploratory proposal to the NSF. The Foundation funded the proposal and eventually granted EDC $4.8 million to develop Man: A Course of Study (MACOS), a year-long program for the fifth and sixth grades using films, tapes, games, and dramatic devices to introduce children to several fundamental questions about human behavior: What is human about human beings? How did we get that way? How can we be made more so?

The program is based on scientific research on animal behavior and on ethnographic studies of human behavior. It focuses in particular on the Netsilik Eskimos, a traditional hunting society from the Pelly Bay region of Canada. It uses these studies to explore fundamental questions about the nature of human beings, their daily life-style, patterns of social interaction, child-rearing practices, and cosmology.[21] The main emphasis in MACOS, however, is ethological, focusing on animal behavior. This orientation was more a political than intellectual decision, a response to the political reality of the educational establishment in which social studies remained dominated by historians. By linking the course to the biological sciences, EDC hoped to avoid interference. This also meant that the course had to be developed at the elementary school level in order to avoid overlapping the new biology curriculum.

Pedagogically, the course is heavily influenced by Bruner's concept of education as a self-motivated process. Bruner argues that the factors leading to satisfaction in learning are curiosity, confidence, identification with the subject, and reciprocity, or the need to respond to others.[22] Children, claims Bruner, need reassurance that they may entertain and express highly subjective ideas; they learn more by creating answers than by finding them in books. The task of education, then, is to provide a stimulating environment that will give children the opportunity to use their own problem-solving skills. Thus MACOS is open-ended and problem-seeking, based on student inquiry and the free exploration of values rather than on teacher authority. Children are asked to compare and contrast what they observe in films and books with their own experience, and to examine their own values as yet another source of data.

MACOS does not avoid controversial issues. The course includes discussions of religion, reproduction, aggression, and murder, for it is built on the assumption that it is necessary to deal with such problems in a thoughtful and reflective manner. "Oversimplification and dogmatism are twin enemies of creative thought."[23] In controversial areas, teachers are advised to encourage their students to cultivate independent attitudes and warned that the questions raised by the course often have no clear-cut answers.

These "liberal pedagogical assumptions" were themselves destined to alienate those textbook critics who were committed to traditional authoritarian relationships in the school system. But the substance of the course and its underlying assumptions were even more controversial. MACOS presents details of animal behavior as provocative metaphors that help to illuminate features of human behavior. The course thus includes a study of the life cycle of salmon, the family life of herring gulls, and the social behavior of baboons, always inviting contrast and comparison with human life.

For example, in the course of discussing animal behavior, the children are encouraged to ask difficult questions about human society. If salmon can survive without parental protection, why cannot man? What differences do parents make? What do you think are the characteristics of successful parents? What is the value of a group to the survival of its individual members? And what is the value of cooperation as opposed to competition? The course thus assumes a discernible continuity between animals and man that remains difficult for many people to accept.

The analysis of Netsilik culture is intended to convey the concept that humanness is an attained and man-made condition "composed of an environment shaped to suit his needs, a society with common rules and expectations, and a spiritual community of mutually held values and beliefs."[24] The study of a distinctive traditional culture that must adapt to a rigorous environment suggests how behavior is shaped by the functional requirements of particular situations. The Netsilik culture offers profound contrasts with American society; it is a society without social hierarchy held together by social bonds and surviving through forms of adaptive behavior uniquely suited to survival in a difficult environment. But an understanding of this culture requires presentation of behavior that seems bizarre and morally repugnant to Western culture (practices of senilicide and infanticide), behavior dictated by the need to minimize the economic burden of maintaining the weak. This of course raises touchy moral questions: "How does our society treat the elderly?" "Do you think your parents ever have to choose between pleasing a friend and doing something they believe is important or right?" "How do cosmologies develop out of culture and experience, and in turn influence this experience?" Teachers are asked to raise these questions without making judgments, and students are asked to explore relationships among their families and friends.

The point emphasized again and again in both animal and Eskimo studies is that neither behavior nor beliefs have absolute value apart from their social or environmental context. "Our hope," claims Bruner, "is to lead the children to understand how man goes about understanding his world, making sense of it; that one kind of explanation is no more human than another."[25] And in Talks to Teachers, MACOS emphasizes that values and moral principles must also adapt to contemporary circumstances: "The rules by which man will have to govern himself must take into account the inventiveness of man himself, the variability of situations in which he finds himself, and the historic process in which change occurs."[26]

MACOS was clearly treading on sensitive ground, dealing with questions that are the foundation of the most dogmatic beliefs. The course is not only built on evolutionary assumptions, but it denies the existence of absolute values, thus explicitly teaching just those controversial ideas that fundamentalists have long suspected were implicit in the teaching of evolution.

Publishers were well attuned to the sensitivity of the marketplace, and the troublesome features of MACOS became apparent when EDC began to seek a commercial publisher. The course had already been successfully tested and evaluated in 300 selected classrooms throughout the country, where it was acclaimed as a major educational contribution.[27] But commercial publishers backed away; they found the need for costly teacher training, the use of expensive audiovisual materials, and above all the new and controversial nature of the course to be significant obstacles. EDC initially sought a joint development-dissemination partnership with a commercial house, and held a bidder's conference to which fifty-four publishers were invited. Twenty-eight attended, and twelve continued their interest, visiting trial classrooms and film showings. These twelve were invited to bid for commercial publication of MACOS, but not one replied. Claimed one publisher, among other problems, "religious groups would not endorse the teaching of this type of material."[28]

EDC continued to search for a publisher, meanwhile developing, with NSF support, a network of regional centers in universities throughout the country. These were staffed by professional educators and scholars in the behavioral sciences, and their role was to disseminate information about the nature of MACOS through lectures and technical assistance to schools adopting the course. By 1969, still with no publisher, EDC sought NSF funds to cover manufacturing and distribution costs. The Foundation agreed to extend EDC a line of credit of $270,000 to back the publication program. With this backing EDC contracted with Curriculum Development Associates (CDA), a small new firm committed to continuing the professional approach to dissemination that had been used in the regional centers. The books were finally published in 1970, and they sold in sufficient quantity that the NSF line of credit was not actually used. NSF phased out its support for regional centers by 1971, but continued to support teacher training programs, contributing in all $2.16 million for MACOS implementation. (See Table 2.)

At the end of 1974, MACOS had been purchased by about 1,700 schools in forty-seven states, and sales were yielding about $700,000 per year. They were to plummet dramatically in 1975 with an epidemic of community disputes.

Table 2. NSF Grant Funds Awarded for Development, Evaluation, and Implementation of MACOS Curriculum (Fiscal Years 1963-1975)

Purpose and Fiscal Year	Awarded
Development	
1963	$ 195,420
1964	513,360
1965	300,000
1966	1,200,000
1967	1,738,000
1968	430,000
1969	270,000
1970	150,000
Total for Development	4,797,380
Evaluation	326,000
History	44,000
Implementation	
1967	35,500
1969	456,000
1970	445,000
1971	387,000
1972	196,000
1973	152,000
1974	285,000
1975	210,000
Total for implementation	2,166,500
Total grant funds	$7,333,880

Source: National Science Foundation Science Curriculum Review Team, *Pre-College Science Curriculum Activities of the NSF, II*, May 1975, p. 92.

Notes

1. For an excellent history of the relationship between educational and social reforms, see Lawrence A. Cremin, *The Transformation of the School* (New York: Alfred A. Knopf, 1962).

2. Hyman G. Rickover, *Education and Freedom* (New York: Dutton, 1959), pp. 189–190.

3. A 1965 survey of a cross-section of grade school teachers found that 92 percent of the 2,000 respondents claimed they would not initiate a discussion of controversial issues in the classroom; 89 percent would not discuss controversial issues, and 79 percent believed they should not be discussed in the classroom at all. See Cremin, *Transformation of the School*.

4. The textbook industry is highly competitive and fragmented. The top ten publishers dominate 50 percent of the sales in the United States, but there are hundreds of smaller publishers who compete for state textbook approval. Gross sales are about $1,550 million annually. Publishers rely on market surveys and records of past sales to estimate demands; they normally write for a national readership, and are not likely to take major risks by developing new material.

5. Congressional support for educational reform in science was evident in NSF appropriations hearings and in the National Defense Education Act of 1958, which subsidized improved teaching in mathematics, science, and foreign languages. See discussion of legislative oversight of early NSF curriculum development in National Science Foundation, Science Curriculum Review Team, *Pre-College Science Curriculum Activities* (Washington, D. C.: U. S. G. P. O., May 1975), vol 2.

6. This program was first called Course Content Improvement Program, then the Curriculum and Instruction Development Program, and later the Materials and Instruction Development Section.

7. The philosophy behind this approach is described by J. J. Schwabb, "Structure of the Disciplines," in *The Structure of Knowledge and the Curriculum*, ed. G. W. Ford and L. Pugno (New York: Rand McNally, 1964), pp. 6–30.

8. NSF, *Pre-College Science Curriculum Activities*, pp. 36–38.

9. By 1973, NSF had supported 7,000 summer institutes, mostly for high school teachers.

10. See Policy Research Division of Library of Congress, *Technical Information for Congress*, Report to the Subcommittee on Science, Research and Development, of the Committee on Science and Astronautics, U. S. House of Representatives, 25 April 1969, ch. 5.

11. *Ibid.*, p. 103.

12. NSF, *Pre-College Science Curriculum Activities*, p. 38.

13. For statistics on dissemination see Joseph Platt, "NSF and Science Education," *Journal of General Education* XXVII (Fall 1975): 188ff.

14. NSF, *Pre-College Science Curriculum Activities*, pp. 63–65.

15. Bentley Glass, "Renascent Biology" in *New Curriculum*, ed. Robert W. Heath (New York: Harper and Row, 1964).

16. Arnold Grobman *et al.*, *BSCS Biology—Implementation in the Schools*, Boulder, Colorado: American Institute of Biological Sciences, Bulletin #3, 1964, p. 1.

17. Addison, Lee, "The BSCS Position on the Teaching of Biology," *BSCS Newsletter* 49 (November 1972), pp. 5–6.

18. The three textbooks, known as the blue, green, and yellow volumes, were published respectively by Houghton Mifflin, Rand McNally, and Harcourt, Brace.

19. BSCS, *Biology Teachers' Handbook* (New York: John Wiley and Sons, 1963).

20. Hillel Black, *The American Schoolbook* (New York: Morrow, 1967).

21. MACOS material includes sixteen films and about thirty booklets, plus games, posters, and records. A complete list is available in a brochure published by Curriculum Development Associates, Washington, D. C., 1972.

22. Jerome Bruner, *Toward a Theory of Instruction* (Cambridge, Mass.: Harvard University Press, 1966).

23. Education Development Center (EDC), ed., *Talks to Teachers* (Cambridge, Mass., EDC 1968), p. 13.

24. Peter Dow, "Man: A Course of Study" in *Talks to Teachers*, ed. EDC, p. 5.

25 Jerome Bruner, MACOS Position Paper, mimeographed (Cambridge, Mass.: EDC, 1964), p. 24.

26. EDC, ed., *Talks to Teachers*, p. 58.

27. In February 1969, the American Educational Publishers Institute and the American Educators Research Association awarded Bruner a prize for his contribution to the instructional process.

28. Peter Dow, "Publish or Perish," mimeographed (Cambridge, Mass.: EDC, May 9, 1975).

II THE SCIENCE TEXTBOOK WATCHERS

I don't know very much, I just know the difference between right and wrong. —George Bernard Shaw, *Major Barbara*

4 Textbook Watchers and Space Age Fundamentalism

Forbidden Subjects

Just as perceptions of social and political needs generate curriculum reform, so social and political tensions provoke educational disputes.[1] Indeed, public schools have been an active political arena in curriculum as well as organizational matters as parents, community groups, or self-appointed textbook watchers contest any change in the substance or methods of teaching that violate their perceptions about the values to be conveyed to future citizens.

Patriotism, educational standards, and morality are three traditional themes of curriculum disputes, emerging in relation to contemporary political and social issues. Waves of immigration, incidents that suggest Communist expansion, or U. S. involvement in international affairs have invariably put textbook watchers on guard against what they feel are threats to American values. Classroom discussions of the United Nations, the labor movement, the New Deal, or socialism have provoked calls for censorship. Textbooks, claimed a patriotic group called America's Future, must "reflect appreciation of American sovereignty and the reasons for preserving it." Schools, after all, could become "a breeding ground for alien ideologies."[2]

These conservative themes emerged just prior to World War II as a reaction to the progressive movement in education. Critics labeled progressive schools "anti-intellectual playhouses" and "crime breeders," run by a "liberal establishment." They wrote about "Treason in the Textbooks," and "Lollypops vs. Learning."[3] Fifteen years later, this sentiment converged with cold war concerns about American technological inadequacy, and the criticism of progressive education grew both more forceful and more widespread. It came from respectable academics as well as from congenital reactionaries. Education was divorced from scholarship, contended the American historian, Arthur Bestor; in democratizing the school system educators had forgotten its function of systematic intellectual training. "Where are our standards?" asked the Council for Basic Education, formed in 1956 to protect educational standards. Critics turned against the professional educators, who were blamed for neglecting scholarship and suspected for their "left-wing liberalism."[4] In the late 1950s, concern over Soviet technological competition reached its peak after Sputnik, and the progressive movement in education collapsed.

Among the critics of progressive education were scientists. Their criticism of the weakness of scientific scholarship in the school system had led to the science curriculum reform movement. But just as scientists associated "technological decadence" with the absence of scientific rationality in education, so textbook watchers would later associate "moral decadence" with the dominance of scientific rationality.

The 1963 Supreme Court decision on the unconstitutionality of forcing children to read prayers in school inspired special vigilance against the influence of secular and scientific values. Rapidly changing life-styles among young people in the 1960s and their rejection of tradition confirmed conservatives' worst fears about the moral implications of declining religious influence.

The decadence of scientism is provoking the resurrection of bacchanals and orgies. One can observe a thirst for ritual . . . the monolithic stranglehold which scientism has on education has issued in the contamination of minds. . . . the erosion of sensibility, the corruption of imagination, and ultimately the explosion of hate.[5]

Thus, by the time the NSF education projects were well underway in public schools, textbook watchers had found another theme—the erosion of religious and moral values implicit in biological and social science courses. The denial that nature was subordinated to a transcendent purpose was, they argued, a primary source of the damaging material culture and the decline in morality that they witnessed.

The popular reaction against scientific rationality in the late 1960s was evident in the wide interest in Eastern mysticism, occultism, astrology, and the pop cosmologies of Velikovsky and Von Daniken. Less visible, but perhaps of more political importance in the long run, was the reaction among an entirely different group of people, expressed in a revival of interest in traditional religion and educational fundamentalism.[6]

The Conservative Ministries

Changing patterns of church membership in the 1960s reflected a renewed interest in conservative religions. Memberships in established Protestant churches began to level off around 1964 and to decline by 1967, but during this same period conservative ministries, often holding very traditional beliefs and making strict demands on their members concerning appropriate behavior, began to grow. Among Lutherans, only the Missouri Synod—the most conservative denomination—expanded its membership. Similarly, the Assemblies of God, the Pentecostal and Holiness groups, Mormons, and Jehovah's Witnesses began to grow steadily around 1960.[7] (See Tables 3 and 4.)

Pentecostal faith healers (such as Oral Roberts) who started their careers in the 1940s began to build universities and Bible institutes. The radio religions expanded to television. One of the most successful of the 1960s preachers was Herbert W. Armstrong, who organized the World Wide Church of God, based in Southern California. Armstrong runs Ambassador College in Pasadena, devoted to restoring the "missing dimension of moral and spiritual values to education." Some fundamentalists associated their movement with the space program. Carl McIntire developed 300 acres at Cape Canaveral as a "freedom center" for his Twentieth Century Reform Movement. He bought the Cape Kennedy Hilton, a convention center, and several other buildings, calling the multimillion dollar complex "Gateway to the Stars, a place that would be witness to His word and a haven for patriots who put their love for God first."[8] Similarly, several astronauts organized High Flight, an evangelical ministry in Colorado.

The reaction against the dominance of scientific values in education was compatible with fundamentalist views. "Our ministry is very proud of the fact that we hold the creationist viewpoint as related in the Book of Genesis," claims Billy James Hargis, minister of Christian Crusade. "We feel strongly that any school that teaches evolution should also teach creation in an equal and fair manner."[9] Herbert Armstrong publishes antievolution material in his magazine *Plain Truth*, which advocates literal belief in "the simple, factual,

Table 3. Membership Statistics for Conservative Churches (1960 and 1970)

Church	Year	Number of Churches	Church Membership
Assemblies of God	1960	8,233	508,602
	1971	8,734	1,064,631
Southern Baptist Convention	1960	32,251	9,731,591
	1970	34,340	11,628,032
United Pentecostal Church, Inc.	1960	1,700	175,000
	1971	2,400	250,000
Seventh-Day Adventists	1960	3,032	317,852
	1970	3,218	420,419
Jehovah's Witnesses (USA)	1960	4,170	250,000
	1970	5,492	388,920
Lutheran Church (Missouri Synod)	1960	5,215	2,391,195
	1970	5,690	2,788,536

Source: National Council of Churches, *Yearbook of American Churches* (Published Annually).

logical record of the Bible." The Armstrong Ambassador College catalogue expresses the following view:

Science, industry, and much of modern education have concentrated on developing the machine rather than the man. The actual bitter fruits of this modern materialist "progress" are increasing unhappiness, discontent, boredom, moral and spiritual decadence . . . the way to peace and happiness is through the Bible . . . the authoritative revelation of the most necessary basic knowledge . . . the foundation of all knowledge and the approach to humanly acquirable knowledge.[10]

This perspective is widely shared in the fundamentalist press, which dwells on "disaster," "lurking evil," "terrible peril," and "threats to survival"—all of these the consequence of scientific secularism.[11]

Fundamentalists criticize scientists for "suppressing" religious views. The Watch Tower Society's Magazine, *Awake* (distribution 7.5 million), has attempted to mobilize its leadership to challenge evolu-

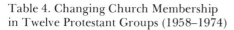

Table 4. Changing Church Membership
in Twelve Protestant Groups (1958–1974)

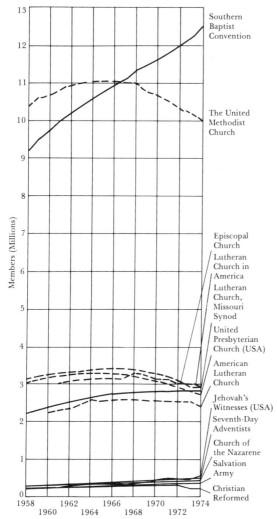

Source: Data compiled from National Council of Churches, *Yearbook of American Churches* (annual statistics). Graph from Dorothy Nelkin, "The Science Textbook Controversies," *Scientific American, 234*, 4, April 1976, p. 36. Note: Membership of groups with a fundamentalist viewpoint (solid line) has increased steadily, while memberhip of nonfundamentalist groups (dotted line) peaked in the mid-1960s and has declined since then.

tionary teaching as "the greatest fairytale ever to masquerade under the name of science."[12] In an article entitled "Do I Have to Believe in Evolution?" *Awake* attacks biologists: "Like a religious hierarchy in the dark ages, they declare *ex cathedra* that evolution is a fact and excommunicate into the outer darkness of ignorance anyone who will not embrace their faith."[13] A set of comic books widely disseminated by Jack Chick Publications in Chino, California, conveys similar themes.[14] *Link Lizard Defeats Evolution* portrays the "superscientist" humbled before evidence provided by Little Laurie, who argues that "Any kind of in-between creatures would be a flop, Doc." A more pernicious booklet, *Big Daddy*, depicts a Jewish-looking professor teaching evolution to his class of long-haired hippies, liberated women, and blacks with Afros. A clean-cut, short-haired blond youth politely and cooly challenges the professor with "facts," citing reputable scientific journals. The professor becomes more and more emotional, trying in vain to support his "beliefs" and finally resigns, defeated and humiliated, as the class turns to religious themes.

In the 1920s, the fundamentalist movement mirrored the discontent of marginal people threatened with disintegration of old lifestyles. It prevailed mostly in small-town rural areas isolated from modern scientific culture. In the 1960s, fundamentalism flourished in the very centers of advanced science and technology, such as urban Texas and southern California—centers of the space industry. Fundamentalists included not only rural people—"the disinherited"—but solid, middle-class suburbanites, often people with technical training. Anxious about the uncertainty caused by rapid social and technological change, they sought new patterns of personal meaning and definitive answers to complex social problems through traditional fundamentalist values.[15]

This renewal of fundamentalism is important to our discussion of science textbook politics because these very conservative people, disillusioned with modernism, are the political and financial base for the movement against science curriculum reform. The fundamentalist ministries are a communications network for the textbook watchers; their interest allows a few very active individuals to wield significant influence.

Some Activists and Organizations

In 1963, Mr. and Mrs. Mel Gabler of Longview, Texas, began to scrutinize the textbooks used in Texas schools. Gabler retired from his job as a clerk in an Exxon district office in 1973 to devote full time to textbook review. Mrs. Gabler is a housewife and mother. The Gablers are among the most active of Texas textbook watchers, appearing at hearings and public meetings to warn communities about threats to their children's education. They have had considerable support from their local home town of Longview, Texas (population: 44,000), which in 1973 awarded Mrs. Gabler a prize as the most outstanding citizen of the year. They have also built up a network of contacts throughout the country among people who share their views.

The Gablers are motivated by concerns with morality. "Have you noticed how texts have been systematically eliminating the basic moral and philosophical precepts of Biblical Christianity and good sense?"[16] They believe that textbooks foster anti-Americanism and hatred for the home and family, as well as the "pernicious idea that man is an animal." The only defense, they argue, is community involvement in textbook decisions. "Unless local people take an active voice in assisting the authorized units of government in the program of selecting textbooks, the selection will continue to deteriorate." They have increasingly turned their attention to the teaching of biology and the social sciences, attacking the evolutionary assumptions that undermine traditional religious beliefs. "Students should be given both sides in order that they may form their own views."

In September 1969, the Gablers' persistent campaign in Texas began to pay off when the board of education removed two of the BSCS textbooks from the state-approved list. This gave them encouragement.

Over the years we have often questioned whether we should continue the textbook work; but this hearing alone justifies the tedious, time-consuming work and budget-drawing expenses for which so few have shown a concern, yet which are vitally important because our nation moves in the exact direction the children are taught.[17]

The Gablers then formed Educational Research Analysts, a nonprofit, tax-exempt organization whose assets are described as "wide experience in textbook evaluation, speakers and materials on request, wide recognition, prayer, support, Christian as well as civic ministry, and gains in the battle for the minds of millions of American students." They have enlisted sympathetic scientists to serve as "expert" witnesses at public hearings: T. G. Barnes, professor of physics at the University of Texas at El Paso, John J. Grebe, a retired chemist from Dow Chemical Company, and Richard LeTourneau, an industrialist and engineer.

In November 1973, the Gablers asked the state board of education to offer courses in both evolution and creation as electives, neither to be taught as part of the required curriculum. Failing this, they proposed guidelines requiring textbooks to identify evolution as only one of several explanations of origins, and to clarify that the treatment is theoretical rather than factually verifiable. This was adopted as an amendment to the Texas Education Policy Act in May 1974. Subsequently, the Gablers took on MACOS as a target and they traveled around the country to organize local protests.

The Gablers and other textbook watchers use the support of the many local organizations that have formed to monitor the educational system. Some of these are branches of the John Birch Society or other national conservative groups that are quite comfortable with the political and religious views of the textbook critics. Some of the groups active in textbook disputes (such as Citizens for Decency through Law) were organized to deal with obscenity issues. The Supreme Court ruling on pornography and obscenity allowed local communities to prevail in setting standards for questionable films. While this ruling was certainly not intended for school textbooks, it helped stimulate demands for local control over educational material which, after all, claimed the textbook watchers, also infringed on local values.

Several organizations watch curriculum innovation because of their concern for the increased influence of the federal government on local educational policies. The inevitable resentment caused by desegregation legislation in the 1960s greatly enhanced the political appeal of this issue of local control, and groups concerned with federal intervention have been extremely receptive to complaints about government influence on curriculum. The Heritage Foundation, for example, is a tax-exempt policy research organization in Washington, D.C., that supports studies expressing a conservative point of view on current issues (federal spending, energy, foreign policy, educational policy). This Foundation is supported at about $500,000 a year by Joseph Coors, a well-known contributor to right-wing causes.[18] Coors believes that government is chipping away at individual initiative through too much interference in the lives of citizens, and he combines this belief with his concerns about "atheism, liberalism, and Godless Communism." Federally suported curriculum is thus a natural target for his Foundation.

Other Washington-based organizations survey the textbook scene to preserve local initiatives in education. Leadership Action, Inc. is a group devoted to "involving citizens in active government participation." This includes increased lay involvement in the selection of textbooks, and decreased federal intervention. NSF-funded courses are one of their targets.

The Council for Basic Education (CBE) was founded in 1956 to preserve the standards of education against the liberalizing trends of progressive education (interestingly, the same issue that motivated the science curriculum reform movement). CBE's original membership of 158 had grown to 4,500 by 1974. CBE has taken an active position against some NSF-supported curriculum, especially concerning the teaching of social science in public schools. It spreads its views through speeches and publications, and by responding to inquiries from educators, journalists, and congressmen. Its director, George Weber, claims that the *CBE Bulletin* is increasingly in demand. "Educators who once asked that it be sent in a plain brown wrapper are now ordering multiple copies."[19]

Weber argues that schools must avoid controversial topics. These include morality, religion, and politics, all of which he sees permeating the NSF social science curriculum. Moreover, he feels that by promoting method (inquiry-oriented instruction) over substance (facts), the federally supported courses perpetuate the same neglect of basic skills that characterized the progressive education movement. The general permissiveness and lack of authority in the new curriculum fosters "rampant illiteracy." "Students are getting diplomas for warming their seats and not striking the teacher."

Political Tactics

The textbook watchers use a variety of political tactics to translate their concerns into administrative decisions; they adjust these tactics according to the educational policy-making structure in different states and to their own access to power within that structure. Where policy decisions concerning textbooks are made by centralized school boards and statewide textbook commissions, these administrative organizations become the target for political pressure. Elsewhere, efforts are directed toward local school boards, and if administrative action seems unlikely to yield change, textbook watchers go to the courts and sometimes to the streets.

In State Legislatures

The repeal of Tennessee's "monkey laws" in 1967 reflected more a change in the alignment of voting districts than a change in attitudes. Antievolution sentiments persisted; in some areas teachers attempting to teach evolution have been reprimanded or dismissed. According to a public opinion poll in September 1972, three quarters of the high school students in Dayton, Tennessee, still believed in creation: "Darwinian evolution breeds corruption, lust, immorality, greed and such acts of criminal depravity as drug addiction, war, and atrocious acts of genocide."[20] And in 1973, less than six years after repealing its antievolution legislation, the Tennessee General Assembly passed a new statute requiring that

Any biology textbook used for teaching in the public schools which expresses an opinion of, or relates to a theory about origins or creation of man and his world shall be prohibited from being used as a textbook in such system unless it specifically states that it is a theory as to the origin and creation of man and his world and is not represented to be scientific fact. Any textbook so used in the public education system which expresses an opinion or relates to a theory or theories shall give in the same textbook and under the same subject commensurate attention to, and an equal amount of emphasis on, the origins and creation of man and his world as the same is recorded in other theories including, but not limited to, the Genesis account in the Bible.[21]

This law, essentially declaring the Bible a reference book for biology, passed the Tennessee House of Representatives by a vote of 69 to 15, and the Senate by 28 to 1.

The National Association of Biology Teachers (NABT) challenged the constitutionality of the legislation, contending in a federal district court that it interfered with free speech, free exercise of religion, and freedom of the press as guaranteed by the First and Fourteenth Amendments.[22] One month later, unknown to the NABT, an organization called America United for the Separation of Church and State, Inc., filed a similar suit in a state chancery court in Nashville. As a result, the district court abstained from considering NABT's suit until the constitutional issues were resolved by the state court. NABT attorney Frederic LeClercq then appealed to the United States Supreme Court, both on the jurisdictional issues and on whether the Tennessee act violated constitutional amendments. The Supreme Court refused to accept the case, but finally on April 10, 1975, a court of appeals in Tennessee overruled the equal time legislation, claiming that it showed

a clearly defined preferential position for the Biblical version of creation as opposed to any account of the development of man based on scientific research and reasoning. For a state to seek to enforce such preference by law is to seek to accomplish the very establishment of religion which the First Amendment to the Constitution of the United States squarely forbids.[23]

This decision was an important precedent, for similar bills had been introduced in the state legislatures of Georgia, Kentucky, Arizona, Michigan, and Washington.[24] Most of these efforts failed, for state legislatures, even if sympathetic in substance, were reluctant to intervene in educational policy making, and textbook critics turned to schoolboards and textbook commissions in their attempts to influence the selection of course materials.

In School Boards and Commissions
In twenty-two states, including Texas and California, there are cen-

tralized state school boards and textbook commissions that make major educational decisions. These are composed of teachers and laymen, often political appointees. The commissions meet every five to six years to select textbooks in various subject areas for the state board of education. While local school districts can use textbooks that do not appear on the list, there are financial incentives to follow state-approved textbooks, for these are usually the only books that are subsidized. Thus it becomes extremely important for publishers to have their books on these lists, especially in the more populous states. Moreover, state recommendations can influence the general policies of textbook publishers, who normally do not print separate editions for each state. A decision in California or Texas may have repercussions throughout the industry, affecting the character of books available in any state. Thus, textbook watchers make a major effort to influence these state-approved lists, and they direct much of their energy toward the state boards of education and their curriculum committees.

They have had some success. Chapter 6 will describe in detail how creationists in California convinced the state textbook commission to propose that creation theory be taught in biology textbooks as an alternative viable theory of origins. While California provided the most dramatic and well-publicized case of political pressure to influence textbook selection, similar activities continue quietly in other states. For example, in September 1973 the Oregon School Board ruled that school libraries must have creationist resource materials and that teachers must encourage students to "weigh the information and arrive at their own conclusions."

Textbook critics have their greatest leverage when they attempt to influence local school districts or even individual schools. In Ohio, for example, an organization called the Creation Research Science Education Foundation, Inc., was formed in 1973 to bring creation material to libraries and schools. This organization established local chapters, called Boosters of True Education, throughout the state to lobby with local school boards. They convinced the Columbus School Board to pass a resolution encouraging teachers to present creation theory along with evolution theory, "so that students may choose."

In most parts of the country, local or community values are reflected in the teaching of science, and the matter never becomes a policy issue. The science coordinator for a school system in southern California organized a service course for high school teachers to explore diverse points of view on questions of evolution vs. creation. "Since scientists do deal with questions of origin, there is a place for creation theory in our curriculum." He hoped there would be a "better balance" in future textbooks. Well aware of current scientific thinking, this man was won over by the textbook critics' argument concerning the "fairness" of presenting alternative beliefs and their plea to respect the values threatened by science education.

But science teachers in high schools in many rural areas simply do not deal with controversial topics. For example, in a high school of Appleton, Wisconsin, topics such as evolution are simply avoided. In this conservative, middle-class community, the home of many fundamentalists, the local people annually decorate the grave of Joseph McCarthy in a formal public ceremony. One of the high school science teachers and at least one of the school board members are "creationists." Teaching the contemporary scientific perspective is hardly a high priority. A high school biology teacher who wanted to present the theory of evolution to his class had to present Genesis as well, allowing students to choose what made most sense to them.[25]

One way in which local textbook watchers deal with educational trends that they find disturbing is to form "fundamental" or "alternative" schools.[26] The alternative school movement first developed in the 1960s, as liberal groups sought less structured educational programs than were offered in the public schools. In the 1970s, however, such schools are being formed by conservatives who resent the "liberalization" of public education. These new "fundamental" schools emphasize basic skills, moral and religious values, and discipline. Their traditional curricula often includes Bible studies and excludes the new math, new science curricula, and other "liberal innovations" that are felt to threaten local values, reduce achievement, or contribute to disciplinary problems. In areas where such efforts cannot be easily organized, textbook watchers bring their concerns to the courts.

In the Courts

In August 1972, William Willoughby, religion editor of the *Washington Evening Star*, filed suit ("in the interest of forty million evangelistic Christians in the United States") against H. Guyford Stever, Director of the National Science Foundation, and the Board of Regents of the University of Colorado. The NSF had supported the development of the BSCS textbooks at Colorado, and Willoughby wanted the Foundation to spend an equal amount "for the promulgation of the creationist theory of the origin of man." Willoughby, who calls himself "a liberal evangelist," claims that citizens are coerced to pay taxes to support educational activities that violate their religious beliefs; supporting educational programs that are "one-sided, biased and damaging" to religious views is an improper use of governmental monies. [27]

Willoughby tries to distinguish himself from other creationists and is "as embarrassed as anyone else" by the excessive zeal of fundamentalists with "hard core, doctrinal demands." He claims his concern is for "impartiality" and "fair play" for those with evangelical convictions, and he resents the "intellectual snobbery" of scientists, who present their ideas as "Truth." He claims to receive many supportive letters from people who are not religious but are concerned with fairness and tolerance for minority beliefs.[28]

Willoughby's formal complaint alleged that the NSF violated the First Amendment: "The government is establishing as the official religion of the United States, secular humanism." The U.S. District Court in Washington, D.C., dismissed the case in May 1973, claiming that the First Amendment does not allow the state to require that teaching be tailored to particular religious beliefs and that the BSCS books were secular. Willoughby went to the U. S. Supreme Court, which eventually dismissed the case in February 1975.

A similar case developed in Kanawha, West Virginia, when the Williams family, assisted by a lawyer from the Heritage Foundation, sued the Kanawha County Board of Education for recommending textbooks that contradicted their religious beliefs by "encouraging disbelief in a supreme being."[29] The family sought an injunction against the offensive material, but the court found no violation of constitutional rights and recommended administrative remedies through the board of education. However, the rural citizens of Kanawha County had already chosen another means of satisfying their demands.

In the Streets
In April 1974, the citizens of Kanawha Country, West Virginia, started a long and violent protest against "godless teaching" in the public schools. The conflict began when the board of education tried to develop a modern English Language Arts Program in the school district of 46,000 students, the largest in the state. This included textbooks identified as "dirty," "antireligious," and generally threatening to local values. During the following months there were pickets, strikes, the closing of mines employing 5,000 workers, shootings, beatings, and the firebombing of school buildings. The civil strife was concerned variously with "dirty books," "secular humanism," and with "federal influence on local education." The militancy was abetted by national organizations such as the Heritage Foundation, whose attorney represented the protesters.

Religious fundamentalists dominate the rural hollows of Kanawha County, the heart of the Appalachian coal district, but if Kanawha is poor, it is by no means the poorest or most isolated part of Appalachia. Less than 25 percent of the rural labor force still works in the mines, and the county has become a commercial and industrial center providing diversified employment and a median income of $7,381 for rural households.[30] Kanawha has experienced rapid social and economic change, and its rural population has not been assimilated into the new social structure. The county is polarized between the relatively sophisticated urban and educated people of Charleston and those in the hollows who resent the domination of the city and take seriously West Virginia's striking motto *Montani Semper Libri*. Fundamentalist religion persists and is even strengthened by having to serve as a cushion for the stresses of social change; indeed the textbook conflict became a religious and cultural war, as parents saw their beliefs ignored by the educated and the powerful urbanites who dominate the school system. The leading textbook critic was the wife of a Church of Christ minister; she confronted a teacher with a doctorate from Yale. As a reporter described the meeting, it was "a war between people who depend on books and people who depend on The Book."[31]

Textbooks teaching evolution were one of many concerns in the Kanawha dispute. Schoolbooks were perceived as advocating sex and crime as well as anti-Christian and anti-American values. The conservative Pentecostal churches, believing in literal interpretation of the Bible, felt their beliefs were violated by both the scientific and the literary texts recommended for schools. They opposed any curriculum that encouraged disbelief. "We don't teach this at home, we don't want this at school."[32]

The Kanawha County Board of Education compromised; it voted to return most books to the classroom but approved a district-wide adoption of creation science textbook material, and stipulated that no student would be required to read any book objectionable to parents on religious or moral grounds. Protest continued, however, with renewed violence. Finally the board adopted new guidelines for textbook selection that excluded many of the disputed books and set up screening committees of laymen to review books for controversial content. Meanwhile, the Kanawha schools reopened, and the most violent protesters were brought to court. But the issues remained unresolved. Should parents or professionals decide on school curriculum? What should be done when modern scholarly thinking infringes on morality as perceived by local communities? Kanawha raised fears of violent resistance to educational decisions—fears that were already an important factor in educational policy following civil rights and busing disruptions. The incident thus led to increased caution concerning the development and distribution of new curricula.

Textbook watchers are antiliberal, often anti-intellectual, and certainly antiestablishment. They are not necessarily antiscience, but rather they object profoundly to what they call "scientism" or "secular humanism," that is, the prevailing dominance of scientific values. Their vision of science is strangely skewed with religious and personal values, yet they will argue the importance of increased public understanding of science in terms that on the surface would be quite acceptable to scientists themselves.

Science has innumerable social implications and applications. Solutions to social problems require real understanding of the origin of the physical processes which affect them (e.g., nuclear energy, fossil fuels, ecology, genetic engineering, hallucinogenic drugs, etc.). Each person needs, more than anything, a sense of his own identity and personal goals, and this is impossible without some sense of his origin. . . . Lack of a sound scientific understanding of origins and meanings among modern young people has impelled them to seek help in such anti-scientific solutions as "mind-expanding" drugs, witchcraft, astrology, and the like.[33]

However, the textbook watchers conclude from this argument that scientific understanding requires a religious perspective; that secular scientific theories based entirely on naturalistic explanations have drastic implications for social and personal behavior. These views are dramatically evident among the "scientific creationists," a group of textbook watchers with scientific training who initiated much of the recent opposition to the teaching of evolution theory.

Notes

1. Mary Anne Raywid, *The Axe-Grinders* (New York: Macmillan, 1962), is a useful history of curriculum disputes. Michael W. Kirst and Decker Walker, "An Analysis of Curriculum Policy-Making," *Review of Educational Research* 41, No 5: 479–501, emphasizes the need to analyze curriculum policy in political terms as "a scene of conflict and uneasy accommodation."

2. Raywid, *Axe-Grinders*, p. 51. See also Jack Nelson and Gene Roberts, Jr., *The Censors and the Schools* (Boston: Little, Brown, 1963).

3. Lawrence A. Cremin, *The Transformation of the Schools* (New York: Alfred A. Knopf, 1962).

4. *Ibid*.

5. Letter to the Editor in the *Medford Mall Tribune*, 15 December 1972.

6. Louis Dupree, "Has the Secularist Crisis Come to an End?," *Listening* 9 (Autumn 1974).

7. Dean M. Kelly, *Why Conservative Churches are Growing* (New York: Harper & Row, 1972). Kelly claims that this growth pattern is due to the rigid commitments required of the membership.
    Kelly's statistics are from the National Council of Churches, *Yearbook of American Churches*. These statistics are questionable, and some studies suggest that much of the impressive increase in membership is due to reaffiliation of people after a peri-

od of inactivity. See R. W. Bibby and M. Brinkerhoff, "Circulation of the Saints: A Study of People who Join Conservative Churches," *Journal for the Scientific Study of Religion* 12 (September 1973): 273–284. Rodney Stark and Charles Glick, *American Piety* (Berkeley: University of California Press, 1968), attribute the growing membership of conservative churches to the social mobility of people from lower classes with fundamentalist leanings into the middle class. They predicted the decline of supernatural and religious faith; as their middle-class status crystallized, members of conservative churches would move into more liberal denominations. This has failed to occur; on the contrary, there is a remarkable increase in conservatism within mainline churches as well as growth among the charismatic and Pentecostal sects.

8. James Morris, *The Preachers* (New York: St. Martins Press, 1973).

9. Billy James Hargis, personal communication.

10. Ambassador College General Catalog, Pasadena, California, 1974–1975 (pp. 21–22).

11. James H. Jauncey, *Science Returns to God*, (Grand Rapids, Mich.: Zondevran Books, 1973), *passim*.

12. Jehovah's Witness, *Awake*, March 1974. Note that the 7.5 million copies are distributed free, often to unwilling recipients. Actual membership in the sect is about 400,000 in the United States, and 2 million worldwide, doubling between 1965 and 1975.

13. *Awake*, September 1974. See also the Jehovah's Witness publication *Did Man Get Here by Evolution or by Creation?* Brooklyn: Watchtower Bible and Tract Society, 1967.

14. Chick runs what he calls a new kind of evangelism based on house-to-house outreach by laymen trained in modern Saul Alinsky-type organizing techniques.

15. An apparent need for more rigid religious programs has begun to permeate the larger, more liberal Protestant denominations as well as the fundamentalist churches. There is increasing criticism of "secular theology" (the reliance on the ability of man to solve his own problems) and an emphasis on "the need for sound fundamental beliefs and personal faith." See, for example, discussion at the 1975 Hartford Theological Foundation.(Conference reported in the *New York Times*, 9 March 1975.)

16. Mel Gabler, "Have You Read your Child's Textbooks?" *Faith*, March/April 1975, p. 10.

17. Educational Research Analysts, *Newsheet*, No. T-110, n.d.

18. See series on Joseph Coors by Stephen Isaacs in the *Washington Post*, 4–7 May, 1975. Coors Beer is one of the largest family-owned firms in the United States, with sales of $440 million in 1973.

19. George Weber, personal interview. The *CBE Bulletin* is published monthly, and has a circulation of about 6,700.

20. *New York Times*, 1 October 1972.

21. Amendment to the Tennessee Code, Annotated Section 49-2008, passed 30 April 1973.

22. There were three coplaintiffs from Tennessee: Joseph Daniels, Jr., and Arthur Jones (professors at the University of Tennessee), and Larry Wilder (a public school teacher). They contended that the legislation violated their academic freedom.

23. National Association of Biology Teachers, *News and Views* XIX (April 1975).

24. The Georgia State Senate had passed equal time legislation in 1973, but this was tabled while The Divine Creation Committee of the Georgia House held hearings. The Committee reported the prevailing public view: that information on all theories of creation should be available in public schools, but recommended that the state board of education deal with the problem through selection of textbooks. Subsequently, the science and creation series was approved for state adoption. In Phoenix, Arizona, creationists backed by the Mormon Church circulated a petition for a referendum to establish legislation that schools teach the Biblical view. They failed to draw sufficient signatures for the referendum, but the same groups later attacked MACOS with much greater success. Textbook watchers in Michigan introduced four bills for "equal time" in the legislature during 1973.

25. The above information on local practices was developed out of personal interviews and correspondence with educators in these areas.

26. The Pasadena School Board, for example, is advocating fundamental schools as a reaction against breakdown in discipline and declining scholastic achievement (*New York Times*, 26 November 1975). See Council for Basic Education, *Bulletin*, for monthly reports on the new alternate school movement.

27. *William Willoughby* v. *H. Guyford Stever*; U.S. District Court for the District of Columbia; brief filed August 7, 1972; civil action 1574–1572.

28. Telephone interview with Willoughby.

29. *Williams* v. *Kanawha Board of Education*, 74-378-CH. (Decision 30 January 1975.)

30. National Education Association, *Kanawha County, West Virginia: A Textbook Study in Cultural Conflict* (Washington, D. C.: NEA Teachers Rights Division, February 1975).

31. Paul Cowan, "Holy War in West Virginia," *The Village Voice*, 9 December 1974.

32. *Washington Post*, 13 September 1974.

33. Institute for Creation Research, "Scientific Creationism for Public Schools," (San Diego, Cal.: ICR: November 1973), Summary.

5 The Scientific Creationists

All religions, nearly all philosophies, and even a part of science tes-
tify to the unwearying, heroic effort of mankind desperately denying its own contingency. . . . The ideas having the highest invading
potential are those that *explain* man by assigning him his place in an
immanent destiny, in whose bosom his anxiety dissolves.[1]

During the 1960s, a group of scientifically trained fundamentalists began to re-evaluate fossil evidence from the perspective of special creation as described in the Biblical record. These creationists,
much like their fundamentalist predecessors in the 1920s, accepted
the Biblical doctrine of creation as literal: "All basic types of living
things, including man, were made by direct creative act of God during the creation week described in *Genesis*."[2] They believe that
creation theory is the most basic of all Christian beliefs, "at the very
center of the warfare . . . against Satan," and the fact that many
churches fail to emphasize special creation is a "tragic oversight that
has resulted in defection . . . to the evolutionary world view, and
then inevitably later to liberalism. . . . "[3] They choose to reinterpret
organic evolution according to Biblical authority.

Some creationists accept aspects of evolution theory but set limits
to scientific explanations, rejecting, for example, natural selection as
a causal explanation of evolutionary change. The more extreme
creationists deny all evolutionary processes, arguing that evolution
and creation are mutually exclusive theories: "You choose to accept
the statements of Scripture, or you choose to accept the claims of
evolutionists. You cannot believe in both."[4] Still others accept the
common compromise—that there are two levels of reality—but they
are concerned that the teaching of evolution denies and obscures all
religious explanation and that failure to teach alternative hypotheses implies that science provides a complete and sufficient understanding of ultimate causes.

Among those who identify themselves as creationists are some fa-
natics (a botanist who claims that evolution theory is "a special argument of the devil"), disciples of traditional fundamentalist sects,

some wealthy industrialists and hotel keepers, several astronauts, and many solid, middle-class, technically trained people working in high technology professions in centers of science-based industry. These modern-day creationists share many of the moral and religious concerns expressed in the twenties, but their style is strikingly different from that of their flamboyant ancestors. Arthur Hays described the circus atmosphere in Dayton, Tennessee, in 1925.

Thither swarmed ballyhoo artists, hotdog vendors, lemonade merchants, preachers, professional atheists, college students, Greenwich Village radicals, out-of-work coal miners, IWW's, single taxers, libertarians, revivalists of all shapes and sects, hinterland soothsayers, holy-rollers, an army of newspaper men, scientists, editors, and lawyers.[5]

In comparison, creationist confrontations are more like debates within professional societies. Indeed, creationists try to present their views at the annual meetings of professional organizations such as the National Association of Biology Teachers. Even during the California public hearings on textbook selection, creationists presented brief technical papers, and the only placard to be seen was a chart of the hydrogen atom intended to demonstrate the scientific validity of creation theory.[6] For creationists argue that Genesis is not religious dogma but an alternative scientific hypothesis capable of evaluation by scientific procedures. They present themselves not as believers but as scientists engaged in a scholarly debate about the methodological validity of two scientific theories.

The Bible as Science

The creationist world view rejects the theory that animals and plants have descended from a single line of ancestors, evolving over billions of years through random mutation. Creationists cannot accept the implication that natural selection is opportunistic and undirected, that selection pressures act to cause genetic change only because of immediate reproductive advantage. According to creation theory, biological life began during a primeval period only five to six thousand years ago when all things were created by God's design into "permanent basic forms." Like the pre-Darwinian Charles Lyell, creationists believe that all subsequent variation has occurred within the genetic limits built into each species by the Creator. Evolution is a directed and purposeful process and present variety among animals is merely part of a blueprint to accomodate a variety of environmental conditions,[7] or "simply an expression of the Creator's desire to show as much beauty of a flower, variety of song in birds, or interesting types of behavior in animals as possible."[8] Change would not modify the original design, for nature is static, secure, and predictable, each species containing its full potentiality.

The creationists thus differ from evolutionists in their explanations of the origin of life, the transmission of characteristics, the nature of variation and complexity, and the character of the fossil record. (See Table 5.)

Clearly, creationists are faced with a formidable amount of evidence that supports the theory of evolution. This poses a cruel dilemma; they must either admit exceptions to their beliefs that would raise doubts among their constituents, or they must maintain consistency at the risk of public ridicule. They have chosen the latter alternative and spend their energies trying to demonstrate that evidence supporting evolution is biased and incomplete, or that it can be reinterpreted to fit whatever conceptual system is convenient. For example, creationist theoreticians argue that the fossil record is far from conclusive, and fails to provide the transitional forms or linkages between diverse living groups that would suggest evolution from a common ancestor. While fossils from the Cambrian period indicate a highly complex form of life, it does not necessarily follow that this life had evolved for over a billion years. There is no fossil record in the Precambrian period to sustain this view.

Creationists also deny the evidence from techniques of radioisotope dating, for these techniques are based on assumptions that no uranium or lead has been lost throughout the years and that the rate at which uranium changes has remained constant over time. Rejecting the uniformitarian hypotheses that allow evolutionists to extrapolate events in the ancient past from present evidence, they argue that if a Supreme Being created the world, and a catastrophe like the Flood altered it, then the evidence for radioisotope dating is simply irrelevant.[9]

Table 5. Alternative Models: Creation vs. Evolution

	Creation Model	Evolution Model
Theory of origins	All living things brought about by the acts of a Creator.	All living things brought about by naturalistic processes due to properties inherent in inanimate matter.
Transmission of characteristics of living things	Creation of basic plant and animal kinds with ordinal characteristics was complete in the first representatives.	All living things originated from a single living source which itself arose from inanimate matter. Origin of each kind from an ancestral form by slow, gradual change.
Variation	Variation and speciation are limited within each kind.	There is unlimited variation. All forms are genetically related.
Complexity	Sudden appearance in great variety of highly complex forms.Net present decrease in complexity.	Gradual change of simplest forms into more and more complex forms.
The fossil record	Sudden appearance of each created kind with ordinal characteristics complete. Sharp boundaries separate major taxonomic groups. No transitional forms between higher categories.	Transitional series linking all categories. No systematic gaps.

Adapted from Henry Morris, "Creation vs. Evolution" *American Biology Teacher*, March 1973.

Creationists also reject the genetic data used to support hypotheses about random mutation, arguing that the same data can be used to deny the theory of evolution; mutations, after all, are usually detrimental and unlikely to contribute to the continuity of life. Similarly, they claim that insights into phylogenetic relationships provided by analysis of protein structure and chromosomal arrangements are based on "dubious assumptions" about the similarity between major plant and animal groups. If one assumed that such groups are unrelated and re-examined the same data according to "polyphylogenetic" assumptions, one would reach quite different conclusions.[10]

For creationists, the law of conservation of energy (that energy cannot be created or destroyed) and the second law of thermodynamics (that energy approaches increasingly random distribution) are additional proof of an initial ordering of natural processes. Similarly, the laws of quantum mechanics—that individual events (the decay of K-mesons) are not predictable—suggest that theories of origin and change are "fundamentally unprovable" but that change is more likely to have occurred by design than by random mutation.[11]

Creationists' "scientific" arguments, which they claim to develop through concurrent studies of scripture and nature, touch on floods, on heredity and genetics, on chemical and radioisotope dating techniques, on the blood circulation system, and on the earth's magnetic field, just to name several of the enormous range of topics covered in their writings. Their "facts" are highly selected and suggest a limited understanding of modern biology and of scientific method.[12] Like Darwin's contemporaries, they view science as an inductive and descriptive process and they seem to comprehend poorly the function of theories and models as useful instruments for prediction. Moreover, when pressed, creationists will argue that design in nature exists simply because of the will of the Creator. They are aware of the problems in this argument, but then they claim that evolution theory is but today's creation myth, based also on faith, although it excludes consideration of a supernatural force. If one accepts a different set of assumptions, then creation theory becomes fully as workable and fruitful a hypothesis as evolution.

Scientists try in vain to refute creationists' arguments. They note the practical and historical problems in a literal interpretation of Genesis and the many facts that contradict creationist theories. They accuse creationists of repeating their arguments, as if repetition could establish verity. Factual arguments and criticism, however, are not likely to change creationist beliefs. Groups committed to particular assumptions tend to suppress dissonant evidence, and criticism only encourages increasing activity in support of existing beliefs.[13] For those who believe in creationism, it is a distinct and coherent logical system that fully explains the world. "Studying the facts of physics and chemistry, I find that the only way I could truly understand the present world is by the word of God and the inspiration of the Holy Spirit."[14] It is evolution theory that is the "scientific fairy tale." While creationists' ideas may appear irrational to a modern biologist, so the ideas of evolutionists are irrational to a creationist. But the accuracy and rationality of creation beliefs are of less interest here than their political efforts to impose these beliefs on the educational establishment. These efforts come from the "research centers" where scientific creationists work.

Creationist Organizations

The American Scientific Affiliation
The American Scientific Affiliation (ASA), formed in 1941, is an "evangelical organization of men and women of science who share a common fidelity to the Word of God and to Christian Faith." Formed mostly by Lutherans who shared concern "over the sweeping tide of scientific materialism and waning faith of modern youth subjected to its influence," the ASA investigates problems bearing upon the relation between Christian faith and science. Most of its 1,750 members carefully distinguish themselves from scientific creationists, whom they would prefer to label "antievolutionists."[15] However, like the scientific creationists, they believe that evolutionary concepts are misleading and have serious moral and social as well as theological implications.

Because of diverse opinions among its membership, the organization has avoided taking a position that advocates teaching creation theory in public schools. ASA does, however, criticize the evolutionary emphasis in textbooks, arguing that evolution is taught in a far too dogmatic way, that the theory is extended beyond what is scientifically appropriate, and that it unnecessarily excludes consideration of alternative theories. "For the great mass of people, acceptance of evolution is not a personal judgment rising from evidence any more than scientific beliefs generally are; rather, popular acceptance represents a deference to scientific authority."[16] Their proposed solution is to compromise; students should learn to appreciate creation and design as alternatives to evolution but not necessarily as part of the biology curriculum. Biology teachers, however, must avoid implying that evolution is the only acceptable theory. This view was to be of considerable political importance, for ASA members have been influential in mediating textbook controversies, especially in California. But their moderation also brought dissension within the organization. Some members argued that the ASA leadership had capitulated to the mainstream of scientific thought. In fact, it was this group of disaffected ASA members who formed the first organization for "scientific creationism."

The Creation Research Society
In 1963, ten men formed the Creation Research Society (CRS) in Orange County, California. They had been members of the ASA, but left that organization when it refused to take a political position on the teaching of evolution. Their objective in founding the CRS was "to reach all people with the vital message of the scientific and historic truth about creation. The Center feels that the false philosophy of evolution is one of the principle roots of all the 'anti-Godisms' that abound today."[17]

The founders of CRS call themselves "scientific creationists." To attain the status of voting membership in the organization, members must meet two requirements: a postgraduate degree in science and belief in the literal truth of the Bible. As of April 1974, there were 514 voting members, plus several thousand associates. About fifty members contribute articles to the organization's journal. With the formation of other creationist organizations in California, CRS moved to Lansing, Michigan, where a branch had developed in the early 1960s. Active Michigan members include a retired chemist from Dow Chemical, several science instructors from Concordia Lutheran College, and a professor of science education at Michigan State University.

CRS promulgates its views in a quarterly journal (circulation 2,000), as well as through notices in various fundamentalist tracts. In addition, CRS supports a stable of speakers ready and willing to lecture on creationism. These activities are supported through membership dues of $7 per year, tax-deductible contributions, and the sale of books and magazines.

Sects concerned with the purity of doctrinal beliefs are characterized by schisms. Creationists are no exception, and CRS soon split into several groups. In 1970, in a struggle over leadership, several members broke away to form the Creation Science Research Center (CSRC) in San Diego, California.

The Creation Science Research Center

CSRC is a small, tax-exempt, research and publishing organization formed "to take advantage of the tremendous opportunity that God has given us. . . . to reach the 63 million children in the United States with the scientific teaching of Biblical creationism."[18] Its research projects, which include investigation of the physical aspects of the Flood, are intended to "clarify problems in the field of geophysics, oceanography and structural geology as well as Biblical and geological chronology."[19] The organization also engages in legal activities to undermine what they claim is illegitimate federal funding of school curriculum. CSRC has accused the California State Board of Education of illegally accepting federal programs, and it has filed injunctions in several California counties to delay the expenditure of funds for "biased" textbooks. Meanwhile, the organization offers publishers its services to "neutralize" textbook material.

CSRC is engaged primarily in its own "curriculum reform program," devoted to the popular dissemination of creationist literature. It publishes a magazine called *Science and Scripture*, a textbook series, film strips and cassettes all colorfully packaged, and an "action kit" including the legal, organizational, and technical information necessary to implement the teaching of creation theory in public schools.[20] Finally, the organization runs a tourist service:

"Visit the mountains of Ararat with the world's foremost authority on the search for Noah's ark—cost, $1,397 from New York."

Distribution of CSRC material is facilitated by the group's association with the Southern California branch of the Bible Science Association which runs a radio ministry and an active extension service. Together the two organizations have a mailing list of about 200,000 individuals and many schools, churches, and textbook committees. In 1972, the CSRC was divided in a conflict over copyright questions and some of its members formed a new organization, the Institute for Creation Research.

The Institute for Creation Research
The Institute for Creation Research (ICR) is the research division of Christian Heritage College, founded in 1970 with the sponsorship of the Scott Memorial Baptist Church, an independent Baptist organization. In 1972, this church purchased a monastery on 30 acres east of San Diego, California, for its parochial high school and shared the site with the college and institute. It is an old, attractive, Spanish-style campus set against the dry but scenic hills of southern California. The corridors of the Spanish-style buildings are bare, decorated with only a few nature scenes. One office door carries the label "Noah's Ark," another, a cartoon: Smidgeon asks a computer, "If you're so smart, how did the world start?" and the computer replies, "Read Chapter 1 of Genesis."

While not an accredited institution, the college, with about two hundred students, aspires to develop full undergraduate and graduate school programs in "the Study of Christian Evidence and Scientific Creationism." The college catalog describes its introductory biology course as "a survey of the life sciences; general and molecular biology; human physiology; creationism in biological origins." Its psychology course includes a section on "the unique nature of man." Its five-man science faculty is involved in an open campaign against evolution theory. All five men are staff members of the ICR.

Creationists are able to sustain their beliefs in the face of considerable criticism because they have strong social support. This is very evident at Christian Heritage College, and at its research institute, where faculty and staff are mostly kin. Tom LaHaye, the college president and pastor of the Baptist church, is a radio minister known in fundamentalist circles for his series of Family Life Seminars, which dwell upon the destruction of family life in modern technological society. His wife, as well as his sister and brother-in-law, works at the College, as do her brother and daughter-in-law. The ICR's senior faculty members are Duane Gish and Henry Morris. Gish's wife is a librarian and four members of the Morris family are also on the ICR staff.

Besides Gish and Morris, three other scientists work on the ICR faculty full time; several others work part time. At least one of these men uses a pseudonym because he worries that identification with creationism will ruin his academic career. However, ICR works to develop a reputation as the "scholarly arm of the creationist movement" debunking the CSRC as "a promotional and sales organization" and its director as a man with "an honorary doctorate from Los Angeles Christian University—a college with no telephone listing and no campus." The use of "false titles," it is feared, reflects badly on the "real scientific creationists." ICR identifies its own primary activity as research devoted to developing an empirical base for creation theory and claims to leave promotional activities to other organizations. Yet, the Institute runs radio programs, conferences, workshops, and summer institutes and publishes a monthly magazine containing both technical articles and current news of creationists' activities. During its first two years the Institute staff published seven books, ran seven summer institutes, delivered some two thousand lectures in thirty-eight states and five foreign countries, and visited about ninety college campuses.

ICR projects include expeditions to find geological evidence that the earth is young and archaeological investigations of Mount Ararat to prove the validity of the Noachian myth. These expeditions are led by John Morris, who is also full-time public relations director for the Institute. There have been many attempts by religious groups to find Noah's Ark, but Morris claims that ICR is the only serious scientific organization that is capable of interpreting the scientific evidence of the Ark's existence. ICR expeditions, however, have been catastrophic. During the first attempt in August 1972, "The men were robbed and beaten by Kurdish outlaws, victimized by city officials, and fired on in an ambush. And three of them were temporarily incapacitated by the body-rending blows of lightening."[21] These events might have been taken as ill omens, but the group planned further trips in 1973, 1974, and 1975, all of which were blocked by the Turkish government.

Despite the small size of each organization and the obvious advantage of coordination, creationist groups are competitive and highly critical of each other. However, their combined activity has brought attention from state legislatures and school boards as well as angry criticism from biologists. News about their activities has appeared in the national media, and publicity, even if negative, encourages creationists to feel that they are finally being taken seriously.[22]

Other Organizations
California is a major center for creationist activity, but similar organizations can be found in other states. The Genesis School of Graduate Studies in Gainesville, Florida, is advertised as the "first known postgraduate level college stressing scientific creationism." It offers a Ph.D. in science-creation research and its curriculum emphasizes "special creation and the young earth model." Its president is the pastor of the Gainesville University Baptist Church. Other Bible schools, such as Bob Jones University in Arkansas, teach courses presenting both evolution theory and special creation.

An occasional creationist course appears in the catalogs of major universities. A professor of chemistry at Southern Illinois University has taught a course called The Creation Alternative, which includes lectures such as "Principles of Geology Revisited." At Michigan State University, creationist John N. Moore teaches a course called Science, Beliefs, and Values, presenting creation theory as an alternative to evolution.

Creationists work with various publishing organizations, such as the Bible Science Association of Caldwell, Idaho, formed by the Lutheran minister Walter Lang, to "set forth the scientific value of the creationists' position." Lang claims that the Bible is scientific and "because of the miracle of inspiration, is also infallible in its scientific statements." This organization distributes the *Bible Science Newsletter* to 27,000 subscribers, sponsors seminars, radio programs, and films "for those who have been misled into accepting evolution theory." The Bible Science Association models itself after a scientific society, holding annual conventions in which speakers are identified primarily by their professional degrees and academic positions. In 1973, a new chapter was organized called the Scientific Creationism Association of Southern New Jersey. Other supportive organizations include Educational Research Analysts in Texas and the Creation Research Science Education Foundation Inc., in Ohio.

A British equivalent, called the Evolution Protest Movement (EPM), founded in 1932, is a "scientific, educational, religious, nonsectarian, non-political and non-profit-making" organization with over eight hundred members devoted to demonstrating that the theory of evolution is not in accordance with fact and causes a decline in morality. EPM has published more than two hundred pamphlets and several books. It was financed by Captain Ackworth, a submarine commander, and its presidents have included Douglas Deward (once an auditor-general of India), Sir J. Ambrose Fleming (the physicist who devised an electron tube in 1904), and Sir Cecil Wakely (a one-time biologist and President of the Royal College of Surgeons).[23] A second British group, The Newton Scientific Organization, was formed in 1973 to distribute creationist literature and to advance the scientific study of creation.

The Activists

Who are these dedicated individuals who have presumed to question one of the scientific community's more strongly held concepts?[24] Henry Morris, director of the ICR and vice-president for academic affairs at Christian Heritage College, has a Ph.D. in hydraulics from the University of Minnesota (1950), and served at one time as a professor of hydraulic engineering and chairman of the department of civil engineering at Virginia Polytechnic Institute (1957–1960). Now in his fifties, Morris has been an active creationist for some thirty years. During his college training at Minnesota, he accepted evolution theory despite his religious upbringing. But during his graduate years, he began to read the Bible and to take a more active role in Christian affairs. "In trying to lead others to Christ, I needed answers and this led me to research. And being an engineer, I looked for solid evidence." Morris had a hard time with his colleagues at VPI, where he claims there are at least thirty-five creationists who are quiet about their beliefs in order to protect their professional standing. Collegial pressures eventually forced him out of the secular university setting. He continues to do some work in applied hydraulics, but most of his work is on "Biblical subjects."

Duane T. Gish, associate director of ICR, received a Ph.D. from the University of California, Berkeley (1953), in biochemistry and held a postdoctoral fellowship at Cornell University Medical School for several years. He spent most of his career as a member of the research staff at Upjohn and Company; he studied scientific evidence related to creation theory as an avocation. In 1971, in his mid-fifties, he began to devote full time to this work, encouraged by the "growing evidence that special creation was valid scientifically." Gish voiced his concern about the "appalling ignorance on the subject of science both within the academic community and among the public." If people understood the nature of scientific evidence, he claimed, they would be more sympathetic to creation theory.

Lane Lester joined the ICR in 1974. He has a Ph.D. in genetics from Purdue University, has taught high school, and was an assistant professor at the University of Tennessee. Lester had been brought up as a devout Southern Baptist. For years he reconciled his religious upbringing with science by believing in theistic evolution (that God set in motion the process of natural selection). Then, in 1972, he heard Gish talk and he plunged into creationist literature, discovering that he could interpret scientific data within a creationist framework. He was gratified that he no longer had to compromise his religious beliefs and joined the CRS. He then took a job at the Biology Science Curriculum Study (BSCS) for a year to learn about methods of developing educational materials—of course, he did not reveal that he was a creationist. In 1974, he left BSCS to join ICR.

Many of the activists in the creationist movement are from the applied physical sciences and engineering. They are mostly people who, like Gish and Lester, once made an uncomfortable accommodation between their religious beliefs and their scientific training. Creationism appealed to them as a means to resolve contradictions.

The creationists claim that applied scientists are interested in creationism because "they have their feet on the ground and are heavily committed to test out theories." Most biologists, they feel, are too "brainwashed" with evolution theory to think flexibly about the evidence. They also argue that people in technical professions, working in highly structured and ordered contexts, are inclined to think in terms of order and design.[25] Another explanation came from Wernher von Braun, the famous NASA rocket engineer, who declared his personal support for the "case for design" as a viable scientific theory: "One cannot be exposed to the law and order of the universe without concluding that there must be design and purpose behind it all . . . I endorse the presentation of alternative theories for the origin of the universe, life, and man in the science classroom."[26]

Clearly, the scientists and engineers active in creation movements are not against science and technology. Many of them earn their living in technical industries. (See Table 6.) Indeed, when questioned on specific contemporary issues, creationist leaders were generally favorable to technology. Some supported development of the SST and the Alaska pipeline. Others expressed ambivalence about the development of civilian nuclear power, but only because the techniques for the disposal of nuclear waste were based on evolutionary assumptions. Far from being against science, the creationists spend much of their energy legitimizing their beliefs in scientific terms, firmly convinced that failure to do so would trivialize them. Their main objection to evolution theory is that it "incorporates all the attributes of a religion"; it is "a doctrine of origin" that replaces God with eternal matter and creation by random mutation; it is a doctrine of salvation not through faith but through foresight and the manipulation of nature. Thus, claim the creationists, it violates traditional religious assumptions and endorses its own system of ethics.[27]

It was in the context of this ambivalence toward science that the new precollege curriculum in biology and the social sciences, both based on evolutionary assumptions, became targets for textbook watchers. In addition to their persistent concern with patriotism and the standards of education, textbook watchers became alerted to re-

Table 6. Advisory Board and Staff of the Institute for Creation Research (1973)

Technical Advisory Board

Name	Discipline, Degree	Present Position, Activities
Thomas G. Barnes	Physicist Ph.D.	Prof. of Physics, U. of Texas, El Paso Pres. of CRS Author of ICR Tech. Monographs Research: Electricity & Magnetism
Edward F. Blick	Nuclear Engineering Ph.D.	Prof. of Aerospace, Mech. & Nuclear Engineeering U. of Oklahoma Research: Fluid Mechanics
David R. Boylan	Engineering Ph.D. (Iowa State)	Dean of College of Engineering Iowa State U.
Larry G. Butler	Biochemist Ph.D. (UCLA)	Prof. of Biochemistry Purdue U.
Kenneth B. Cumming	Biologist Ph.D. (Harvard)	Dept. of Biology, U. of Wisconsin, LaCrosse Research: Fishery Biology
Malcolm A. Cutchins	Engineering Mechanics Ph.D. (VPI)	Prof. of Aerospace Engineering Auburn U., Alabama
Donald D. Hamann	Food Technology Ph.D. (VPI)	Prof. of Food Technology N. Carolina State U.
Charles W. Harrison	Electrical Engineering & Applied Physics Ph.D. (Harvard)	Sandia Laboratories Albuquerque, NM
Harold R. Henry	Fluid Mechanics Ph.D. (Columbia)	Chairman, Dept. Civil & Mining Engineering U. of Alabama
Joseph L. Henson	Entomology Ph.D. (Clemson)	Chairman, Science Division Bob Jones U.
John L. Meyer	Zoology Ph.D. (Iowa State)	Asst. Prof. of Physiology & Biophysics U. of Louisville Medical School
John N. Moore	Science Education Ed.D. (Mich. State)	Prof. of Natural Science Michigan State

Table 6 (continued)

Technical Advisory Board

Name	Discipline, Degree	Present Position, Activities
Charles C. Ryrie	Theologian Th.D. (Dallas Theol. Seminary) Ph.D. (U. of Edinburgh)	Prof. of Systematic Theology Dallas Theological Seminary
John C. Whitcomb, Jr.	Old Testament Scholar Th.D. (Grace Theol. Seminary)	Prof. of Theology, Dir. of Post-Grad. Studies Grace Theological Seminary
Duane T. Gish	Biochemistry Ph.D. (Berkeley)	Prof. Natural Science Christian Heritage College Assoc. Dir. of ICR
Henry M. Morris	Hydraulics Ph.D. (U. of Minn.)	Dir. ICR VP Acad. Affairs Christian Heritage College
Harold S. Slusher	Geophysics M. S. (U. of Okla.)	Dir. Planetary Science Program Christian Heritage College
Robert Franks	M. D. (UCLA)	Prof. of Biological Science Christian Heritage College
William A. Beckman	Chemistry Ph.D. (Western) Reserve College)	Prof. Physical Science Christian Heritage College
Stuart Nevins	Geology M. S. (San Jose State)	Prof. Earth Science Christian Heritage College
Maurice Nelles	Engineering Physics Ph.D. (Harvard)	Lecturer Christian Heritage College
Lane Lester*	Genetics Ph.D. (Purdue)	Lecturer Christian Heritage College

*Arrived in 1974.
Source: *Acts and Facts*, Nov.–Dec., 1973

ligious and moral issues as creationists organized to fight what they perceived to be efforts to "indoctrinate" their children with the dangerous values of "secular humanism."

Notes

1. Jacques Monod, *Chance and Necessity* (New York: Vintage Books, 1972), pp. 44, 166.

2. Creation Research Society brochure.

3. Richard Bube, "Science Teaching in California," *The Reformed Journal*, April 1973, pp. 3–4.

4. Joseph L. Henson, "Theistic Evolution," *Faith*, March/April 1973, p. 23.

5. Arthur Garfield Hays, "The Scopes Trial" in *Evolution and Religion*, ed. Gail Kennedy (New York, D. C. Heath, 1957), p. 35.

6. The ironies of the California Hearings were described by Calvin Trillin, "U.S. Journal: Sacramento, California," *The New Yorker*, 6 January 1973, pp. 55ff.

7. John N. Moore and Harold Slusher, *Biology: A Search for Order in Complexity* (Grand Rapids: Zondervan, 1970), p. 422.

8. *Ibid*; see review by Wyatt Anderson, Rossiter Crozier, and Ronald Simpson, *The Georgia Science Teacher 13* (1974): 15–18.

9. Duane Gish, "Creation, Evolution, and the Historical Evidence," *The American Biology Teacher* 35 (March 1973): 23–27.

10. John N. Moore, "Evolution, Creation and the Scientific Method," *The American Biology Teacher* 35, January 1973, pp. 23–27.

11. Ronald S. Remmel, "Randomness in Quantum Mechanics and its Implications for Evolutionary Theory," Testimony to the California State Board of Education, 19 November 1972.

12. In fact, their efforts to "prove" the existence of God by scientific reasoning is also a curiosity even according to conventional Christian scholars, who argue that the existence of God can only be experienced or believed—not "proved."

13. See the work on cognitive dissonance by Leon Festinger, *A Theory of Cognitive Dissonance* (Evanston, Ill.: Row Peterson, 1957). The relationship between beliefs and the interpretation of scientific information is discussed by S. B. Barnes, "On the Reception of Scientific Beliefs" in *Sociology of Science*, ed. Barry Barnes (Harmondsworth: Penguin Books, 1972), pp. 269–291.

14. Letter in *Acts and Facts*, November/December 1973.

15. Some scientific creationists have, however, retained their ASA membership.

16. Carl F. H. Henry, "Theology and Evolution," in *Evolution and Christian Thought*

Today, ed. Russell L. Mixter, (Grand Rapids, Mich.: Eardmans Publishers, 1959), p. 202.

17. Creation Research Society brochure.

18. In 1974 the CSRC employed eighteen people and used twelve outside technical consultants. They claimed to have over 10,000 regular donors who could be counted on for small gifts at every fundraising appeal.
Quoted material is from Creation Science Research Center *Report*, October 1973.

19. Creation Science Research Center brochure.

20. CSRC claims that the science and creation textbook series sold about 30,000 copies to each grade level and that one of their most popular books sold 70,000 hardbacks.

21. *San Diego Union*, 23 February 1974, p. B-8.

22. Creationists were interviewed on the "Today" show on April 15, 1974.

23. The Evolution Protest Movement has branches in England, Australia, New Zealand, Canada, and Johannesberg, South Africa. It has published several hundred pamphlets intended to demonstrate that the theory of evolution contradicts scientific fact; titles include "Ape Men are Fakes or Fiction," "Man from Monkey Myth," "How the British Museum Chooses our Ancestors," "God is Science Plus," "Evolution: the Great Delusion," "An Atheist Kicks Against the Pricks," "God Took Risks in Making Men and Monkeys," and "Evolutionary Magic or Creative Miracle." See *Nature* 241 (February 1974): 360.

24. Most of the following information was obtained from personal interviews, from brochures published by creationist organizations, and from curriculum vitas.

25. A study of the religious orientation of scientists finds that those in applied fields who work outside major universities are more orthodox in their religious beliefs than other scientists. Chemical engineers, for example, are much more likely than other scientists to attend church regularly and to avow their belief in life after death. This persists regardless of educational level. Ted R. Vaughan, et al., "The Religious Commitment of Natural Scientists," *Social Forces* 44 (June 1966): 519–526.

26. Wernher von Braun, Letter to John Ford, published in *Science and Scripture*, March/April 1973, p 4. von Braun later qualified his position, stating that he believed there was "divine intent" behind the processes of nature, but did not believe that all living species were created in their final form 5,000 years ago. Several of the astronauts, who surrounded their NASA space program with religious ritual, have also endorsed the creationist view. For example, the astronaut, James Irwin, is a creationist. After his experience on the moon ("I feel the power of God as I'd never felt it before"), he founded an evangelical foundation called High Flight. (*New York Times*, 26 April 1974, p. 18C.) Astronauts Frank Borman and Edgar Mitchell have also indicated that they feel the Genesis account of creation to be an appropriate explanation. Mitchell wrote to Vernon Grose that "I strongly favor the presentation of both points of view with the added hope that such duality will ultimately lead to one of two eventualities. First, the scientific community may modify our model of

living organisms . . . The second alternative might be that scientists will postulate a new unified field concept that will allow predictive incorporation of a distinct, energetic mechanism which interacts with the fields of matter that produce a higher order functioning of life which we call 'consciousness'—a view distinctly different from that of a chance origin in evolution of life." (Letter, 16 June 1972.)

27. Institute for Creation Research, *Acts and Facts*, June 1975.

III TEXTBOOK DISPUTES

Socrates: Did you say you believe in the separation of church and state?

Bryan: I did. It is a fundamental principle.

Socrates: Is the right of the majority to rule a fundamental principle?

Bryan: It is.

Socrates: Is freedom of thought a fundamental principle, Mr. Jefferson?

Jefferson: It is.

Socrates: Well, how would you gentlemen compose your fundamental principles, if a majority, exercising its fundamental right to rule, ordained that only Buddhism should be taught in public schools?

Bryan: I'd move to a Christian country.

Jefferson: I'd exercise the sacred right of revolution. What would you do, Socrates?

Socrates: I'd re-examine my fundamental principles.

—Walter Lippmann, *Four Dialogues*

6 Creation vs. Evolution: The California Controversy

In 1963, two women from Orange County, Nell Segraves and Jean Sumrall, decided to "seek justice for the Christian child." They justified their demand on the basis of the 1963 Supreme Court decision (*Abington School District* v. *Schempp*) that it was unconstitutional to force nonbelieving children to read prayers in school. Segraves and Sumrall argued that if it is unconstitutional "to teach God in the school," it is equally unconstitutional "to teach the absence of God." They cited the majority opinion of Justice Clark: "We agree, of course, that the state may not establish a 'religion of secularism' in the sense of affirmatively opposing or showing hostility to religion, thus 'preferring those who believe in no religion over those who do believe.'"[1]

In 1963, assisted by Walter Lammertz, a geneticist and one of the founders of the Creation Research Society, they petitioned the state board of education to require that textbooks clearly specify that evolution is a theory rather than truth. Arguing that Christian children must have equal protection under the law, they sought a legal opinion from the Department of Justice concerning the teaching of theories believed by atheists. If religious persons cannot teach their doctrines in public schools, why should atheists and agnostics be allowed to do so? Assistant Attorney General Norbert A. Schlei agreed that it would be unconstitutional for a state to prescribe atheism, agnosticism, or irreligious teaching. Max Rafferty, then California Superintendent of Public Instruction, promptly ruled that all California texts dealing with evolution must clearly label evolution as a theory.

Rafferty was notoriously sympathetic to conservative and fundamentalist causes. In a California Department of Education booklet called *Guidelines for Moral Instruction in California Schools*, he unequivocally expressed his concern with "protecting the child's morality from attack by secular humanists."

I always think that America was built on the Bible . . . The teaching of evolution as a part of the religion of Humanism, therefore, is yet another area of concern . . . If the origins of man were taught from the point of view of both evolution and creation, the purpose of education would be satisfied.[2]

In 1966, Rafferty encouraged creationists to demand that creation theory be given equal time in biology classes, claiming it was consistent with the education code of the 1964 Civil Rights Act prohibiting teaching that reflects adversely on any persons because of race, color, or creed. This code states that references to religion are not prohibited as long as they do not constitute instruction in religious principles. Creationists interpreted this as a sanction for the teaching of creation theory as an alternate scientific hypothesis.

The state board of education denied the creationists' 1966 proposal for equal time. Then, in 1969, the California State Advisory Committee on Science Education prepared a set of curriculum guidelines for its public school science programs called *The Science Framework for California Schools*, intended to be the model for science curriculum development in California.[3] The Committee is an advisory group appointed by the board of education. Its fourteen members consist of scientists and teachers, and they are advised by distinguished consultants. Their advice is implemented by the State Curriculum Commission, a group of sixteen laymen and scientists, thirteen of whom are appointed by the board of education and one each by the California Assembly, Senate, and Governor. This commission advises the board on specific textbooks that are appropriate within the guidelines of the *Science Framework*. Textbook decisions are made every six years. Local school districts are not required to abide by state selections, but they lose state subsidies for textbooks outside the state-approved list. The state department of education works closely with publishers to make any changes that are required by the curriculum commission.

Creationist Demands

In October 1969, the state advisory committee presented a draft of the *Science Framework* to the board of education. It contained two paragraphs on evolution, and several of the nine members of the board objected. These included Howard Day, president of the board and a Mormon; John Ford, M.D., a physician and Seventh Day Adventist; David Hubbard, President of Fuller Theological Seminary in Pasadena; Thomas Harward, M.D., a Mormon and personal physician to Max Rafferty; and Eugene Ragle, a Baptist.[4] Their objections were described in the *Los Angeles Times* and read by Vernon Grose, an aerospace engineer and expert on systems safety who works for a consulting firm in Los Angeles. Grose belongs to the American Scientific Affiliation and to the Assemblies of God, a Pentecostal denomination. He is a "commission-sitter" serving on fourteen state commissions, and a writer of numerous archconservative tracts on the decline of American morality. He was "called" to action by the discussion of evolution theory in the *Science Framework* and decided to fight the evolutionary bias that was "threatening our national heritage."[5]

Can you imagine the impact on the logic required for justice in our courts if we were forced to amend the Declaration of Independence to read: we hold these truths to be self-evident, that all men arose as equals from a soup of amino-like molecules, and that they, by virtue of this common molecular ancestry, are endowed with certain inalienable rights. . . . [6]

On November 13, 1969, Grose presented a thirteen-page memorandum to the board of education, arguing that the theory of creation be included in textbooks as an alternative explanation for the origin of life. He reduced this to a brief statement for the *Science Framework*.

All scientific evidence to date concerning the origin of life implies at least a dualism or the necessity to use several theories to fully explain relationships, . . . While the Bible and other philosophical treatises also mention creation, science has independently postulated the various theories of creation. Therefore, creation in scientific terms is not a religious or philosophical belief. Also note that creation and evolutionary theories are not necessarily mutual exclusives. Some of the scientific data (e.g. the regular absence of transitional forms) may be best explained by a creation theory, while other data (e.g., transmutation of species) substantiate a process of evolution. . . .[7]

The board of education unanimously accepted the statement and printed it in the *Science Framework*. Thus, forty-five years after the Scopes trial, the guidelines for a state educational system that serves one million children included a formal recommendation to teach creation theory.

The California State Advisory Committee on Science was horrified by the revision. "The changes, though small in extent, have the effect of entirely undercutting the thrust of the 205-page document . . . offend[ing] the very essence of science, if not religion."[8] The committee publicly repudiated the document, and the panel of science advisors appointed by the curriculum commission to advise on the choice of textbooks resigned.

The implications began to be evident in 1971, when the curriculum commission selected specific biology textbooks to be used in the schools. No creationist texts were among the books submitted to the board of education. Dr. Ford, vice-president of the board, reminded his colleagues on the commission that "No textbook should be considered for adoption . . . that has not clearly discussed at least two major contrasting theories of origin."[9] In May 1972, the board restored the omitted texts and reorganized the commission, changing its name to the Curriculum Development and Supplemental Materials Commission. Creationists, including Vernon Grose, were well represented; the new commission included only one professional scientist, Junji Kumamoto, a chemist, who for several years was to engage in a one-man defense of evolution in California.

California has about one million children of school age and buys ten percent of the nation's textbooks. Grose was responsible for negotiating with publishers, and in June 1972, he called publisher's representatives to ask them how they intended to include creation theory. He found some quite willing to adapt to the new *Science Framework*. One proposed replacing a section about Leakey's archaeological discoveries of primitive man with a reproduction of Michaelangelo's Sistine Chapel painting of the Creation and a drawing of Moses. Another submitted a fourth-grade science text that claimed that science had nothing to say about who made the world and why. One chapter had as an exercise an investigation of the Biblical account of creation.[10] One fifth-grade text on "concepts in science" mentioned that George Darwin was the son of a famous

English scientist, Charles Darwin; this is the book's only reference to Darwin.[11]

To prepare for the final adoption of the textbooks, the board of education called a hearing on November 9, 1972, in order to assess public opinion. The hearing became a confrontation between creationists and evolutionists. It promised to be a circus—Dayton 1972—but bureaucratic procedures (five-minute limitations on speeches) and the creationists' efforts to present themselves as scientists set a tone of sober debate. Engineers appointed to curriculum development commissions somehow lack the fire of fundamentalist preachers. Yet the ironies were striking. "Witnesses from each side appeared in each other's clothing," observed a journalist amused by the spectacle of scientists speaking for creation theory and theologians supporting science.[12] The twenty-three witnesses for Genesis included only three Baptist ministers, but twelve scientists and engineers. The evolutionists, on the other hand, called forth only four scientists. Other witnesses included Presbyterian, Episcopalian, and Mormon ministers, Catholic and Buddhist priests, and a rabbi; all testified for the need to separate science and religion.

Evolutionist Response

Evolutionists were incredulous that creationists could have any influence. "It just does not make sense in this day and age." Incredulity led to amused disdain. The British journal, *Nature*, confidently offered free subscriptions to the first ten biologists who could claim that their present observations are inconsistent with the commonly accepted views of evolution.[13] A Stanford biochemist placed the creationists' argument "in the same arena as those advanced by the Flat Earth Society."[14] Facetious remarks were abundant. It was proposed that Bible publishers insert a sentence in Genesis to indicate that "scientific method rejects the supernatural approach to explaining the universe."[15] A biologist and member of the state advisory committee inquired whether a scientific course on reproduction should mention the stork theory.[16]

John A. Moore, a biologist (who, to his dismay, is often mistaken for creationist John N. Moore), satirized creation theory by examining it critically as a serious scientific hypothesis. In a masterful exercise in Biblical exegesis that reminds one of Darrow's cross-examination of Bryan, he pointed out problems of accuracy revealed by scholarly disagreement about different versions of the Scripture. He noted internal contradictions in the account of creation in Genesis and practical impossibilities involved in literal interpretation of the Bible. How can one explain in rigorous scientific terms the practical difficulties involved in the Noachian myth: the migration of animals, the necessary size of the Ark, the coexistence of species?[17]

The creationists' public claims of scientific verity were especially embarrassing to biologists. Until recently, scientists have rarely aired their disputes in public. Mindful of their public image and eager to avoid political interference, they usually try to avoid public exposés of arguments among themselves. Control is maintained through informal internal communications and through a peer review system that determines research funding and the acceptance of papers by journals. Creationists, however, claimed publicly to be scientists, and they adopted the language and forms of science. Yet, by seeking external political approval of the validity and justice of their arguments, they ignored the constraints imposed by the norms of the scientific community.

As creationists persisted in their efforts to influence textbook selection' the biologists' amusement and disdain gave way to defensiveness. If creation theory was placed on an equal footing with Darwinism, it would further confuse school children's understanding of what science was about. Thus, scientists countered creationists' demands with legal and political strategies, and they attempted to discredit the movement by refusing to acknowledge the creationists' claim to scientific status.

The National Association of Biology Teachers (NABT) organized the political and legal opposition. NABT is a national organization devoted to the improvement of biology teaching. It has about 8,000 members, mostly high school and junior college teachers, and its journal, *The American Biology Teacher*, is distributed to about 13,000 subscribers. Increasingly dismayed by the California events, in March 1972 NABT organized a committee to plan legal action and retained legal counsel in California, hoping to prevent the board of education from implementing the *Science Framework*. During the following spring and summer, NABT tried to arouse the interest of the scientific community. It set up a Fund for Freedom in Science Teaching, receiving contributions of about $12,000 to support its legal and organizational activities, and it organized a response from professional societies.

The prestigious National Academy of Sciences was moved for the first time to interfere in an issue involving a state decision. In October 1972, the Academy issued a strongly worded resolution.

Whereas the essential procedural foundations of science exclude appeal to supernatural causes as a concept not susceptible to validation by objective criteria; and

Whereas religion and science are, therefore, separate and mutually exclusive realms of human thought whose presentation in the same context leads to misunderstanding of both scientific theory and religious belief; and

Whereas, further, the proposed action would almost certainly impair the proper segregation of the teaching and understanding of science and religion nationwide, therefore

We . . . urge that textbooks of the sciences, utilized in the public schools of the nation, be limited to the exposition of scientific matter.[18]

The American Association for the Advancement of Science also "vigorously opposed" the inclusion of creation theory in science textbooks.

Scientists have built up the body of knowledge known as the biological theory of origin and evolution of life. There is no currently accepted alternative to scientific theory to explain the phenomena. The various accounts of creation that are part of the religion and heritage of many people are not scientific statements or theories. They have no place in the domain of science and should not be regarded as reasonable alternatives to scientific explanation for the origin and evolution of life.[19]

The American Anthropological Association was less vigorous, for some members argued that it would be absurd to offer any response to the creationists' demands. The Association eventually urged legislative and administrative bodies "to reject all efforts for the compulsory introduction . . . of statements reflecting religious and philosophical beliefs subject to different orders of verification into biology textbooks."[20]

The Academic Senate of the University of California condemned the creationists' statement in the *Science Framework* as a "gross misunderstanding" of the nature of scientific inquiry. Finally, nineteen California Nobel Laureate scientists petitioned the state board to leave the teaching of evolution, the only theory based on scientific evidence, intact.[21]

Biologists called the arguments for design based on the intricacy and beauty of nature "spurious and irrelevant." Belief in an intelligent designer is "as blasphemous as it is far fetched."[22] Admitting that much evidence for evolution theory is circumstantial and incomplete, they defended evolution theory as a useful model with solid support from the sum of evidence accumulated in such diverse disciplines as genetics and biochemistry.[23] Like any theory it is based on inferences that are deeply embedded in the structure of the discipline and taken for granted in the course of day-to-day research. The important point for biologists is that research consistently confirms the evolutionary hypothesis and that it remains a powerful predictive instrument, whether or not the evidence supporting it is complete.

The creationists responded to these arguments by asserting that Biblical authority presents an equally useful model that is also confirmed by evidence. During public hearings on the California textbooks, each group claimed the other based its beliefs on faith; each group argued with passion for its own dispassionate objectivity; and each group brought its social and political concerns to the discussion of science curriculum. Scientists and creationists alike bemoaned the moral, political, and legal implications of the alternative ideology. The influence of alternative assumptions on religious equality, as well as on educational practice, concerned both groups. And as each side defended its position and criticized the other, their arguments were strikingly similar. (See Table 7.) Indeed, the debate assumed the aspects of a battle between two dogmatic groups, as the anti-

Table 7. Contrasting Arguments of Creationists and Evolutionists*

Creationist Argument	Evolutionist Argument
On Scientific Methodology	
Creation theory is as likely a scientific hypothesis as evolution. Neither theory can be supported by observable events, neither can be tested scientifically to predict the outcome of future phenomena, nor are they capable of falsification. Evolutionists, while claiming to be scientific, confuse theory and fact. And it is unscientific to present evolution as a self-evident truth when it is based on unproven *a priori* faith in a chain of natural causes. Based on circumstantial evidence, evolution theory is not useful as a basis for prediction. It is rather, "a hallowed religious dogma that must be defended by censorship of contrary arguments." The situation is a trial of Galileo in reverse.[1]	Creationism is a "gross perversion of scientific theory." Scientific theory is derived from a vast mass of data and hypotheses, consistently analyzed; creation theory is "Godgiven and unquestioned," based on an *a priori* commitment to a six-day creation. Creationists ignore the interplay between fact and theory, eagerly searching for facts to buttress their beliefs. Creationism cannot be submitted to independent testing and has no predictive value, for it is a belief system that must be accepted on faith.
On Moral Implications	
Man is a higher form of life made in the image of God. To emphasize the genetic similarity between animals and man is socially dangerous, encouraging animal-like behavior. As a "religion of relativism," evolution theory denies that there are absolute standards of justice and truth and this has disastrous moral implications.	"Tampering with science education by insisting on the priority of feeling over reason, of spontaneity over discipline, of irrationality over objectivity, the honorable man wrecks his own ideals. By attempting to redefine science for his own purposes, the honorable man finds himself in the company of a young hippie radical representing the counter-culture, who indiscriminately is throwing out a life of reason based on objectivity and thus gives himself license to live carelessly and dogmatically."[2]
On Political Implications	
Evolution is a scientific justification for "harmful" political changes. The evolutionary philosophy, which substitutes concepts of progress for the "dignity of man" has been responsible for "some of the crudest class, race and nationalistic myths of all times: the Nazi notion of master race; the Marxist hatred for the	Creationism is a form of right-wing conservatism, as evident in the role of Reagan appointees in the California Board of Education. "Attempts to legislate belief systems through controlling printed materials in the public schools have frequently been a part of fascism."[4]

Table 7. (continued)

Creationist Argument	Evolutionist Argument
On Political Implications	
bourgeoisie; and the tyrannical subordination of the worth of the individual to the state."[3]	
On Legal Implications[5]	
Public schools cannot legally deal with questions of origin that are the domain of religion. They infringe on constitutional rights as guaranteed under the "establishment of religion" and "free exercise" clauses of the first amendment, the "equal protection" and due process clauses of the fourteenth, and constitutional guarantees of freedom of speech. Teaching evolution amounts to the establishment of "secular religion," interfering with free exercise of fundamentalists' truths and violating parental rights. Moreover, restricting the teaching of alternate theories violates the free speech right of teachers.	Exclusion of creation theory in science classes is justified by the first and fourteenth amendments. It is unconstitutional to teach children in a way that would blur the distinction between church and state. Creationism is nonscientific and religious, and therefore to include it would amount to "the establishment of religion." Imposing nonscientific demands would restrict the freedom of teachers to teach and students to learn. To require equal time for doctrines that have no relation to the discipline of biology would impose unconstitutional constraints on the teacher's freedom of speech.
On Religious Equality	
To select one set of beliefs over another is to suggest that one group of people is superior to another. Creationists are a persecuted minority. In view of the wide range of beliefs among Americans, teaching evolution is divisive and inequitable and refects the dogmatism of an established group. "Science has been oversold in Western culture as the sole repository of objective truth . . . the authoritarianism of the medieval church has been replaced by the authoritarianism of rationalistic materialism."[6]	Creationist demands that their beliefs be taught in public schools represent the tyranny of a minority; a few people are using democratic protections to subvert majority interests. To teach creation theory would violate the beliefs of other religious groups. Justice, in this case, would also require teaching hundreds of other mythologies reflecting the belief of the American Indians, Hindus, Buddhists, Moslems, and so on. Religions can co-exist with science because they operate at a different level of reality.
On Sound Educational Practice	
Education in biology is "indoctrination in a religion of secular humanism." It is a breach of academic freedom to prevent the teaching of arguments that	To include creation theory in scientific classes would be poor pedagogy leading to ridicule and rejection of both science and religion. If creation were presented

Table 7. (continued)

Creationist Argument	Evolutionist Argument
On Sound Educational Practice	
have withstood challenges for 6,000 years. "Science demands that our children be taught an unproven, undocumentable theory. There is neither a scientific nor moral base upon which to refuse our school children access to another, much documented theory—the theory of *Genesis* creation." Sound educational practice requires teaching creation as an alternate theory so that students can decide what to believe for themselves.	as an alternate hypothesis, even less would be taught about science than is taught today. Furthermore, it would be a breach of academic freedom to require to teach what is essentially a belief system. In-depth studies of the relationship between science and religion are too sophisticated for public schools. It is sound educational practice to focus on an accurate presentation of scientific fact and leave the teaching of religion to the home.

*Sources: From statements at the *Public Hearings on California Biology Textbooks*, Sacramento, 9 November 1972, and from the *BSCS Newsletter*, November 1972, unless otherwise indicated

[1] John N. Moore, "Evaluation, Creation and the Scientific Method, *American Biology Teacher*, 35 (January 1973); and editorial, *Christianity Today*, 17 (January 1973).

[2] David Ost, "Statement," *American Biology Teacher*, 34 (October 1972), p. 414.

[3] Carl Henry, "Theology and Evolution" in R. Mixter (ed.), *Evolution and Christian Thought Today* (Grand Rapids: Eardmans, 1959), p. 218.

[4] Ost, *op. cit.*

[5] Discussion of the legal issues appears in F. S. LeClercq, "The Monkey Laws and the Public Schools: A Second Consumption?" *Vanderbilt Law Review*, 27 (March 1974), p. 209–42.

[6] Duane Gish, "Creation, Evolution and the Historical Evidence," *American Biology Teacher*, 35 (March 1973), p. 140.

dogmatic norms of science faded with the effort to convey the validity of the theory of evolution.

Despite the formal response from scientific societies and a general dismay at the revival of creationism, many individual scientists were extremely reluctant to become involved. Some were sensitive to political accusations from schoolboards and preferred to stay insulated from the controversy in order to avoid interference in their own work. Others had never been interested in political issues and were simply uncomfortable with any public activity such as talking to reporters, writing letters to editors, or appearing at hearings. Some were disturbed at the idea of opposing a minority group. One biologist, for example, called the NABT fund "the fund for suppression of incorrect theories." Although far from believing in creation theory, this scientist had no desire to outlaw a controversial point of view from the public school system. "Darwin by this time is quite immune to overthrow from fundamentalist attacks, and it isn't going to harm a single child to be made aware that there are divergent opinions available on the subject." He supported an open comparison of creationist and Darwinian arguments. "I am not in favor of the suppression of dissenting opinions."[24]

NABT soon discovered that its own membership included creationists. Letters poured into the Washington, D.C., office. "I do not support your editorial position and the vicious scientific attacks on the creationists." "I feel the fund is being misused to try to force everyone into one mold. It is worse to block false approaches than to tolerate them." One writer suggested that the fund would be used to "promote atheism and agnosticism in the schools." "It should be called a HUSH fund ('Help us to Silence Him')." "It is a campaign to close the mouths of those who espouse theories other than those of evolution."

NABT also found itself caught in the middle of a debate over editorial policy. Should its journal consider publishing creationist articles? Should it print letters from creationists? For several years, the journal had included occasional creationist articles qualified to indicate that they did not reflect the NABT view. The editor felt that their bias and lack of logic would be obvious and self-defeating. In 1972, this editorial policy received a deluge of criticism from scientists, who complained about the journal's lack of discretion. "Creationists' goals are obviously to discredit science and scientists." The journal should not present such "trash." By November 1972, as the California situation reached its climax, the journal stopped publishing creationist articles, although it continued to include a few letters.

A similar discussion took place when NABT organized its 1972 convention. Should creationists be allowed to hold a formal session at this meeting? Some scientists felt that it was necessary and appropriate to include creationists. Biologist Claude Welch was dismayed at the "persecution" of creationists, whom he felt would only make fools of themselves. "Creationists will prosper and multiply on martyrdom, but will perish on exposure. Are we so insecure? Will we let our exasperation with the creationists' irrationality provoke us to become irrational ourselves?"[25] Welch felt the controversy excelled as "soil for nurturing the education process," an opportunity to clarify evolution to an uninformed public. From a different perspective, an NABT regional director felt that creationism should be discussed as an important social issue that bears on science teaching.[26] Other scientists, however, strongly opposed a creationist panel; to allow creationists a voice at the Association's meeting would imply some acceptance of their legitimacy. William Mayer of the BSCS, the science curriculum most vulnerable to creationist attacks, argued that creationists were using the NABT meeting as a stage on which to present their religious ideas. He condemned them as "religious missionaries, concerned primarily with converting classrooms by . . . smuggling religious dogma into classrooms in a Trojan horse."[27] Mayer criticized NABT as "schizoid," fighting the inclusion of creationists' material through legal means while at the same time providing them a forum at meetings.

Creationists eventually held their panel, and 1,500 biology teachers attended. Later, BSCS itself would be criticized as having "sold out" to creationists, after it developed a sound-slide program called, An Inquiry into the Origin of Man: Science and Religion. Intended to "end the debate that has engaged the Western world for well over a century," the slides present cosmological sciences from Hindu, Hebrew, Greek, and Roman myths. It then describes evolution theory, emphasizing the conflict with religion. Biologists quickly accused

BSCS of trying to profit from the conflict; the film, they felt, would only prove to creationists that biologists were concerned with religious ideas after all.

After November 1972, biologists were increasingly reluctant to acknowledge the creationist movement. The 1973 NABT convention did not mention the controversy, and biologists consistently refused to debate creationists in hopes of avoiding any activity that would suggest the scientific legitimacy of the movement. Many tried to discredit the movement by questioning the credentials and competence of those who claimed to be scientists. They were, claimed the biologists, only "engineers." "They are educated at Bible colleges." "Who are these people?" "They are false authorities." "Dullards." "Rejects from the space age." "Is it legal to misuse professional titles?" "Creationists get their doctorates in a box of crackerjacks." "It is a publishers' racket." "As phony as a $3 bill." "A way to subsidize religion." Biologists attacked textbook publishers for responding to such pressure groups. "We should have a fund for protection of prostitution of publishers."

While biologists groped for an appropriate response, creationists merely interpreted their opposition as a sign of recognition and their reluctance to debate as proof that few biologists were willing to defend the theory of evolution and expose its shortcomings in a public forum.

The California Solution

Educational policy-making bodies are well-accustomed to responding to political pressure. Criteria for the selection of textbooks are continually being adjusted to meet the demands of the times. A state board's responsiveness depends on the pressure that various groups can bring to bear on board members. For example, the concerns of ethnic and racial minorities and of women were a source of many textbook changes during the 1960s.

The California education establishment was especially sensitive to issues concerning the teaching of biology, for there had been long legislative debates about the use of live animals for scientific experiments. These had culminated in an ambiguous amendment to the

State Education Code limiting the use of vertebrate animals in classrooms, and many science educators, worried that the life sciences instructional program was jeopardized, were already active in educational politics during the early 1970s.[28] Thus the California State Board of Education was deluged with letters, resolutions, and petitions from educators and scientists, and soon after the November textbook hearings, it began to re-examine its policies. In December 1972, the curriculum committee announced to the board that its members had agreed unanimously on guidelines that would ensure the neutrality of science textbooks. They proposed to eliminate all scientific dogmatism by changing offending statements in textbooks to indicate their conditional nature. Books should discuss *how* things occur and avoid questions of ultimate causes; unresolved questions should be presented as such to students, with an emphasis on the tentative nature of evolution theory. The board of education accepted these recommendations, voting 7–1 to treat evolution as a speculative theory, a decision described in the *Los Angeles Times* as "A Victory for Adam & Eve." The board appointed a committee to implement the recommendations; two of its members, Richard Bube and Robert Fischer, were members of the American Scientific Affiliation; the other two were Dr. John Ford and David Hubbard, state board of education members who had revised the *Science Framework*. None of these men were biologists, and all identified themselves as creationists who accepted the teaching of evolution *if* it remained neutral on the subject of ultimate causes. Bube, for example, defined the issue as "neither creation vs. evolution, nor design vs. chance, but the existence of a Supreme Being, an issue that lies beyond the limits of science."[29] The committee edited the textbooks to clarify both the potentialities and the limitations of science, intending to guard against the "religion of science" as well as "other" religious positions.

The committee prepared a statement, to be printed in all textbooks dealing with evolution, that science cannot answer questions about "where the first matter and energy come from," for scientific methods can only deal with the "physical mechanisms involved." The statement declares that, while the term *evolution* can be used to describe observable processes, the accuracy of the theory of evolution in reconstructing life in the past "depends largely upon the validity of the assumptions on which it is based."

The committee screened thirty textbooks, proposing many changes, and taking particular care to replace specific words that implied acceptance of evolution theory. They changed "developed" or "evolved" to "appeared" and "unfolded" to "occurred." They deleted some words ("ancestors," "descendents," "origins"), and they added qualifying phrases (". . . according to one particular point of view"; "It is believed, in the theory of evolution. . . ."; "The evidence is not clear, but. . . ."). Pictures were relabeled: "This is an artist's conception of what might have been," or "Some people think that plants might have looked like this: What do you think?" And they prefaced each section discussing evolution with a statement indicating that "science has no way of knowing how life began." Some specific changes are listed in Table 8.

Most changes were basically unobjectionable and, indeed, a few did correct some unnecessarily dogmatic statements. The board accepted the revision committee's recommended changes by a vote of 5–3. Since the publishers submitted their bids for basic texts after the *Science Framework* was published, the revisions were considered "technical," that is, based on criteria agreed upon prior to the contract, and they were forced to comply.

Scientists accepted the changes with relief. While they felt the qualifications did injustice to the great body of scientific expertise, the changes were far less disturbing than anticipated. Furthermore, two biologists, Garrett Hardin and Barbara Hopper, wrote a new page 106 for the *Science Framework*, and their revision, focusing on evolution, was accepted by the board of education in March 1974.

The scientific creationists felt "sold out" by what they viewed as minimal changes that neglected their demand for equal time—but by no means did they give up. They began to gather survey material to prove the extent of public support for their views. The Seventh Day Adventist Church of Crescent City' California, polled 1,500 adults, about 57 percent of whom attended church. They claimed that 91 percent of church attenders and 85 percent of nonchurch attenders favored teaching creation in the public schools. Fifty-four percent of church attenders and 65 percent of nonattenders also favored the teaching of evolution. They respectfully submitted their findings in support of "equal time" to the board of education as a form of "public service" in order to help the board "represent our community." Creationists continued their speaking engagements, particularly in California, and they remained highly visible, as newspapers played up their colorful speeches. On college campuses, they often received a favorable press that stimulated months of controversy in the student newspapers.[30]

In February 1973, the creationists presented a new motion for "equal time." They nearly won, for the board upheld its earlier decision by only a one-vote margin. However, a motion by John Ford to place creation in social science books passed unanimously, and a social science framework committee was formed in April to develop the necessary guidelines required by the board.[31] The committee's proposed framework provided that "various views of human origins together with various approaches to the relationship of religious beliefs to scientific theory, must be seen as part of the intellectual and cultural diversity of our society."[32]

The creationists continued to gather evidence of public support, polling the Cupertino Union school district near San Jose' California, one of the largest elementary school districts in the state. A survey of a random sample of 2,000 residents, conducted by Citizens for Scientific Creation, asked, "Should scientific evidence of creation be presented in the schools along with evolution?" The survey found that 84.3 percent of the respondents agreed. The poll also questioned respondents about their personal convictions, to find that 44.3 percent believed in creation, 23.3 percent in evolution, 18.3 percent were undecided, 10.6 percent believed in neither, and 3.5 percent in both. Among those claiming to be evolutionists, 75 percent favored the inclusion of both theories, evidently influenced by the assertion that scientific evidence for creation does in fact exist.

Despite the polls and a swarm of letters and petitions to the state board of education, creationists began to lose their influence in 1974, when the board initiated a new method of evaluating educational materials that would include many more civic organizations and lay-interest groups in the textbook evaluation committee. Creation science books were eliminated, as this broader participatory

Table 8. Changes in Biology Textbooks Recommended by the California Board of Education

Changes in Definitions

Original Version	Changed Version
Science is the total knowledge of facts and principles that govern our lives, the world, and everything in it, and the universe of which the world is just a part.	Science is one way of discovering and interpreting the facts and principles that govern our lives, the world and everything in it, and the universe of which the world is just a part. The scientific way limits itself to natural causes and to descriptions that can be contradicted, at least in principle, by experimental investigation.
Evolution is a central explanatory hypothesis in the biological sciences. Students who have taken a biology course without learning about evolution probably have not been adequately or honestly educated.	Evolution is a central explanatory hypothesis in the biological sciences. Therefore, students need some knowledge of its assumptions and basic concepts.

Qualifications to "Reduce Dogmatism"

Original Version	Changed Version
Scientists can reconstruct the (prehistoric) animal.	Scientists do their best to reconstruct the (prehistoric) animal.
Scientists believe that these species were ancestors.	According to the evolutionary view, these species were ancestors.
A short description . . .	A short approximate description . . .
Modern animals that are descendents . . .	Modern animals that seem to be direct descendents . . .
Some fish began to change.	Some fish began to change although we don't know why.
The earth had spore-bearing plants long before the first seed plants appeared.	There is evidence that the earth had spore-bearing green plants long before the first seed plants appeared.
Paleontologists . . . have reconstructed past history . . .	Paleontologists . . . have done their best to reconstruct the past history . . .
How do we know . . .	On what basis has it been concluded that . . .

Table 8. (continued)

Qualifications to "Reduce Dogmatism"	
Original Version	Changed Version
The evidence that shows how . . .	Evidence that is often interpreted to mean . . .
They would have to change.	They would have to be different.
Fishes adapted . . .	Fishes were adapted . . .
Paleontologists have been able to date the geological history of North America.	Paleontologists have assembled a tentative outline of the geological history of North America.

Changes to Avoid Evolutionary Assumptions	
Original Version	Changed Version
Slowly, over millions of years, the dinosaurs died out.	Slowly, the dinosaurs died out.
As reptiles evolved from fishlike ancestors, they developed a thicker scaly surface.	If reptiles evolved from fishlike ancestors, as proposed in the theory of evolution, they must have developed a thick scaly surface.
Many scientists believe that the universe had a beginning similar to that of a snow fort. They believe that the stars and the galaxies of the entire universe in the beginning were in the form of very small scattered particles.	Science, by definition, cannot say anything about where the first matter and energy of which our universe is made came from. That is because there cannot be any science without matter and energy to deal with. When scientists speak of the beginning of the universe, therefore, they mean the first interactions of matter and energy. Many scientists believe that these first interactions were like those in making a snow fort . . .
Shortly after the flying reptiles took to the air, the early birds developed.	Birds appear in the fossil record shortly after flying reptiles.
The constant rate at which radioactive elements give off particles enables scientists to determine how long it will take for one ball of any sample of a radioactive element to form another element.	Scientists know that radioactive elements give off particles today at a constant rate. If they assume that this rate has remained constant back in time to the date of interest, they can determine how long a . . .

Table 8. (continued)

Changes to Avoid Evolutionary Assumptions	
Original Version	Changed Version
Scientists believe life may have begun from amino acids or viruses, neither of which is usually considered living. Scientists believe life may have been transported from another planet.	Scientists do not know how life began on earth. Some suggest that life began from non-living material. Others suggest that life may have been transported . . .
Plants took to the land and conquered it.	Plants appeared on the land.

Source: From a notebook distributed by the California Board of Education containing single pages from textbooks with changes typed on slips of paper that were superimposed on the original passages.

base gave extremist positions less influence. On May 8, 1974, the board reversed the decision to include creation theory in social science classes. The vote of 5–5 left creationists complaining bitterly about injustice and educational tyranny. "The public schools are not controlled by the public nor does the public have any say in the educational process. It seems that the public through taxes simply pays for that which they do not want."[33]

The creationists were, however, able to delay the publication and distribution of the social science framework until January 1976. The final version did not require discussion of human origins in social studies classrooms or the presentation of creation as an alternative to evolution, but the board of education did advise that both evolution and creation theory be discussed as examples of the intellectual and cultural diversity of society. As a further concession to creationists, the board sent a memo to school districts to remind teachers that whenever human origins were discussed, alternative theories should be presented.

Indefatigable, the creationists accelerated their campaign in order to influence the 1976 revisions of the *Science Framework*. Once again they argued that "theistic and materialistic philosophies are equally religious and/or anti-religious and . . . equally inaccessible to falsification by experimental test." Therefore, they concluded, "both should be studied in the light of the scientific data."[34] (See Appendix 3.)

Meanwhile, textbook watchers were expressing their concerns with science education in other ways. A vulnerable target had appeared in a new NSF social science curriculum that explicitly raised several controversial issues only implied by the evolutionary assumptions of biology textbooks.

Notes

1. Nell Segraves and Jean Sumrall, *A Legal Premise for Moral and Spiritual Guidelines for California Public Schools* (San Diego: Creation Science Research Center, n.d.).

2. Max Rafferty, *Guidelines for Moral Instruction in California, A Report accepted by the State Board of Education* (Sacramento: California State Department of Education, May 1969), pp. 7, 64.

3. For a review of the work of this committee and the background to the creationist controversy, see John A. Moore, "Creationism in California," *Daedalus* (Summer 1974): 173–190.

4. Board of Education members were appointed by Governor Ronald Reagan, and the selection clearly reflected his personal piety.

5. Grose's motivation and his qualifications are suggested by his own statement: "My citizenship really is in heaven. And even though I wasn't trained in biology, when I got into the issue I believe I must have felt something like Jesus did when he over threw the tables and the money changers in the temple . . . the odds were extremely high against success. Yet I believe, because my trust was in the Lord, and because the issue was a significant one, that He honored the effort." Cited in Frederic S. LeClercq, "The Monkey Laws and the Public Schools: A Second Consumption?" *Vanderbilt Law Review* 27 (March 1974): 242.

6. Vernon Grose, statement to California Board of Education, 1969 (mimeographed).

7. California State Department of Education, *Science Framework for California Public Schools* (Sacramento, 1970), p. 106.

8. Paul DeHart Hurd, a spokesman for the committee, cited by Walter G. Peter III, "Fundamental Scientists Oppose Darwinian Evolution," *Bioscience* 20 (July 1970): 1069.

9. Board of Education, minutes of meeting, 8 July 1971.

10. The texts considered were a creationist book by Leswing, *Science, Environment and Man*, and a Macmillan science series. Both were later omitted from the California textbooks screening process.

11. John E. Summers, M.D., "Letter to Editor," *Science*, 9 November 1973, p. 535.

12. Nicholas Wade, "Creationists and Evolutionists: Confrontation in California," *Science*, 17 November 1972, pp. 724–729.

13. *Nature* 239 (October 1972): 420. Two scientists immediately responded to the challenge.

14. David S. Hogness, Statement at a meeting of the California State Board of Education, 6 November 1972 (mimeographed).

15. John E. Summers, M.D., Statement at hearing of the California State Board of Education, 9 November 1972 (mimeographed).

16. Ralph Gerard, Statement at hearing of the California State Board of Education, 9 November 1972 (mimeographed).

17. John A. Moore, "On Giving Time to the Teaching of Evolution and Creation," *Perspectives in Biology and Medicine* 18 (Spring 1975): 405–417. Delivered at AAAS annual meeting, San Francisco, March 1974.

18. National Academy of Sciences Resolution, 17 October 1972. Note that this was approved by only thirty-five of sixty members attending the October meeting.

19. Resolution printed in the *BSCS Newsletter*, November 1972.

20. *BSCS Newsletter*, November 1972.

21. *BSCS Newsletter*, November 1972.

22. G. Ledyard Stebbins, "The Evolution of Design," *The American Biology Teacher* 35 (February 1973): 58.

23. For discussion of the nature of proof and explanation in biology, see Michael Ruse, *The Philosophy of Biology* (London: Hutchinson, 1973); *BSCS Newsletter* 49 (November 1972): 3.

24. These quotes and the following are from correspondence to the NABT in late 1972.

25. Memo to NABT, 28 November 1972.

26. Letter from Wendell McBurney to Jerry Lightner, 22 December 1972.

27. *American Biology Teachers Journal*, April 1974, p. 246.

28. Clifford Frederickson, "Use of Live Vertebrate Animals in Science Instruction in California Schools," *California Science Teachers Journal*, October 1974. The amendment was an addition to the California Education Code, Article 2, Section 1, cl. 8 in Senate Bill 112, signed 1 June 1973.

29. Richard Bube, "Science Teaching in California," *The Reformed Journal*, April 1973, p. 4.

30. See exchange of letters in the UCSD, *Triton Times*, 16, 23, 30 January; 2, 6, 20 February; and 19 March 1973.

31. A social science education framework was developed that required analysis of belief systems to include discussion of creation theory in which "teachers assist students to see that there is evidence for the evolutionary theory and for each of the theories or myths of creation."

32. Quoted in *San Diego Union*, 14 April 1974.

33. This is from a report mailed to 75,000 people from the director of CSRC (Kelly Segraves) in May 1974.

34. Robert E. Kofahl, "Position Paper: The Science Framework Should be Revised With Respect to Evolution and Creation," submitted to California Science Committee of Curriculum Development and Supplemental Materials Commission, by CSRC, October 1975. (See Appendix 2.)

7 The Proper Study of Mankind. . . . The MACOS Dispute

Education is a major industry in Corinth, a small city in an agricultural region of upstate New York.[1] Corinth is both a university town and a commercial center for the surrounding region, and it has experienced typical "town-gown" strains. The Corinth school district was one of the earliest in the country to adopt MACOS and, not surprisingly, the first school to request the course was Lakeview, located in the upper middle-class university suburb where most of the students were faculty offspring. MACOS was implemented at Lakeview School just as intended by its developers; a university team of educators and scientists arrived to train teachers to use the materials. The principal informed parents about the course prior to its implementation. It was enthusiastically received.

In September 1973, the principal of the Springbrook School in Corinth also ordered MACOS for its fourth-, fifth-, and sixth-grade classes. The Springbrook School, less than a mile away from Lakeview, is in a quite different neighborhood; its residents are employed in local civil service and technical jobs, and few have more than a high school education. In implementing MACOS the school administration did not consult with parents or seek local approval. The two teachers responsible for MACOS received no special training; and they were not warned that the course material might be controversial.

For four months, there was, in fact, no hint of a problem. One teacher was enthusiastic about the pedagogical value of MACOS; the other was ambivalent, concerned about the lack of traditional factual and historical content. In January 1974, one of the teachers discussed some of the evolutionary assumptions of MACOS. Knowing the conservative religious feelings of some of her students, she carefully prefaced her discussion with remarks on the existence of diverse beliefs on the question of origins. But the discussion led to an argument, and several children brought to school religious tracts that specifically condemned evolution theory. The children also brought MACOS booklets home, and in a few days a group of mothers "stormed the school." They came to observe classrooms, to question students and teachers, and they argued that the school must give equal time to traditional religious beliefs if it was "preaching" evolution.

"We're all animals, kids are taught here"

One hundred and twelve students were taking the MACOS course in Springbrook School; the parents of twenty-five of these children protested. The principal set up a series of meetings with parents to explain the course and its intentions, hoping that parental concerns could be aired in a rational context. By this time, however, opinions were well formed, and the conflict escalated; according to one of the teachers, "Parents were rude; they questioned my credentials during class period, leafed through the materials on my desk, and recorded my replies to their questions." They also petitioned the board of education to have the principal fired. While the discussion of evolution triggered the protest, other issues soon entered:

We're all animals, kids are taught here.

The children are thinking too much about values; they are too young to be exposed to this type of thing.

Their morals must be shaped at home.

If we cannot sing Christmas carols, then we are not going to let you teach other kinds of religion.[2]

Parents complained about the "permissive" format of MACOS, and suspected that their children were missing traditional subjects. Above all, the community resented "indoctrination by outside experts," and, reflecting local town-gown tensions, they associated the course with "those experts on the campus." Why, asked the parents, were they not consulted before introducing a course that was so offensive to family values?

MACOS had supporters in the neighborhood, but few cared to get involved in the public controversy. The newspaper printed Letters to the Editor from parents and children who liked the course, but they all came from other school districts. Many Springbrook parents voiced their approval of MACOS in private conversation with the principal but hesitated to express their opinion publicly. Or else they felt that their voice did not count, a lack of confidence sustained by the schools' initial neglect of parental opinion. In addition, there was some fear, triggered by telephone threats, that open support of MACOS might endanger their children.

The anti-MACOS groups were well organized. Four parents from a local organization called Save our Schools, concerned with censoring "dirty books in the classroom," served as spokesmen. About seventy-five people from the community regularly attended meetings to discuss how to get rid of the program. Some participants from local fundamentalist churches focused on religious issues, using passages from the Bible to support their opposition. Others had overt political concerns about the "Communist" values they perceived in MACOS. Some resources for organizing the protest came from outside the community; local parents had access to duplicating facilities and to Xeroxed materials from the national media. They were knowledgeable about anti-MACOS protests elsewhere. During the dispute, several speakers from the John Birch Society in a nearby city came to town to talk about "educational issues," and focused on MACOS.

The Springbrook principal sought to resolve the problem by allowing students to substitute another course in American history. Their parents wrote the following excuses to permit their children to withdraw from MACOS: four wanted a more traditional curriculum; eight claimed the religious and moral content of MACOS threatened their personal values; six thought their children too immature to cope with the content; one was concerned with peer pressure; two claimed their children did not want the course; and one simply claimed to be confused about the issue.

The new course began in February, but opposition continued at great cost to the students and staff. Students took sides in the controversy, and many found it difficult to handle the conflicting attitudes of their parents, their teachers, and their peers. Some children who themselves enjoyed the course were placed in the difficult position of defending their parents' objections against the authority of the school. In February 1974, at the height of the dispute, the sixth-grade class took their annual Stanford Binet IQ test. The results were dismal; the previous year the same group of children had scored significantly higher. Convinced that the anxiety over MACOS had influenced the children's performance, the principal administered a different form of the same test three months later when the tension over MACOS was over. Allowing for differences in age and time spent in school, the scores improved by more than 30 percent. (See Table 9.)

The debate was also costly in terms of staff time. During the first three months of 1974, the principal devoted at least one staff meeting per week to the MACOS issue and held eight evening meetings with parents. Finally, in April 1974, the Springbrook School dropped MACOS entirely. This, claimed the principal, was a political, not a curriculum, decision; the tension simply had to be dissipated.

Table 9. Changes in Average IQ Scores of a Sixth-Grade Class During a MACOS Dispute

	May 1973	February 1974	June 1974
Reading	41.0	32.5	42.2
Math	43.7	35.3	48.6

The Politics of Local Protest

MACOS had been acclaimed by teachers, parents, and students throughout the country as a major innovation in science teaching.[3] Yet the Corinth controversy was only one of many; protests spread like an epidemic in 1974, and sales of the curriculum declined by about 70 percent. (See Table 10.) One of the first MACOS protests took place in Lake City, Florida, in 1970 when the course was being tested. Integration decisions in Florida had left the population tense and suspicious of any changes in educational policy. The liberal assumptions underlying MACOS were an obvious target, and the course could not be taught in Lake City schools.

One year later parents in Phoenix, Arizona, raised such strong ob-

Table 10. Man: A Course of Study Domestic Sales 1971–1975 (in rounded figures)

Annual Totals		
1971	$563,000	
1972	602,000	
1973	494,000	
1974	677,000	
1975	205,000	
Quarterly Totals	1974	1975
First	$ 31,000	$35,000
Second	245,000	55,000
Third	291,000	48,000
Fourth	110,000	67,000

Source: Curriculum Development Associates
Note the marked decline in sales beginning in the second quarter of 1975 reflecting the publicity from congressional hearings.

jections to MACOS that the state superintendent of public instruction banned purchase of the material throughout the state until the case was settled.[4] Subsequent controversies erupted in New York, Vermont, Tennessee, Florida, Oregon, Alaska, Maryland, Pennsylvania, Idaho, California, and Texas.[5]

Some selected quotations from MACOS critics during these controversies suggest the range of concerns over the course.[6]

1. Teaching that Man is an Animal Violates Religious Beliefs
"I will never say I came from an ape."
"I do believe it should be countered or balanced with alternative theories of the origin of life, and obviously what I am getting at is Genesis."
"Teaching that man is an animal and nothing more is denying the existence of God and religion. Should such teaching be banned in the public schools as unconstitutional?"
"Children are warned to distinguish between human-like behavior and attributing human motives to animals. It seems that children at that age are not able to make the distinction and do tend to overdo the similarity between animals and man."
"I wonder how many parents would be happy to see their son identify with a baboon instead of his father."

2. MACOS Teaches Disturbing Values
"It teaches that violence and power are necessary for survival."
"It will break down the moral fiber of American youth."
"The course is a steady diet of bloodletting and promiscuity."
"It is violent, unnecessary, and even heretical. The study of Eskimo infanticide has serious implications for abortion practices by students expected to feel that it is up to some authority who or when a baby becomes human."
"It is part of a humanistic cult that claims there is no God or creative force within the universe."
"It perpetuates the humanist idea of one world."
"It openly favors a communal way of life."
"It is used for the political subversion of our children."

3. Educational Experts Undermine Parental Authority
"It alienates the beliefs, values, and allegiances of children, alienating them from their parents."
"The education experts are dictating our values."
"St. Jerome (Bruner) is trying to indoctrinate students."
"The course should be brought up as an election issue or at least judged by a lay curriculum committee."

4. Substitutes for Traditional Education Are Not Useful
"There are, after all, no facts."
"Does cultural relativism make better citizens?"
"How does it prepare students for their future experience?"
"It will take the heart out of education."
"The mood of the country is strongly to return to basic education."
These or similar objections were repeated in every community conflict, but the emphasis shifted somewhat in each, depending on immediate local concerns. Thus, in the university community of Corinth the town-gown conflict assumed importance, as local groups targeted "those experts on the campus." Religious issues dominated Bible Belt protests; cultural relativism was a concern in isolated rural communities. Complaints about the moral implications of MACOS prevailed in urban disputes. Aspects of the course thus became targets for the expression of local frustrations.
MACOS protests often appeared as isolated incidents, but there

was considerable communication among the activists in various disputes. The same few individuals, well-known textbook watchers, appeared at the site of many controversies; for example, the Gablers, who had been monitoring textbooks in Texas, traveled around the country to win their "battle for the minds of American students." The Gablers were instrumental in removing BSCS textbooks from the Texas Board of Education's approved curriculum in 1969, and they have since carried on a continued diatribe against evolution theory and smut. MACOS was on their disapproved list and they lectured in many of the communities where protests developed. Ona Lee McGraw of Maryland, an officer in the National Coalition for Children and a member of Leadership Action, Inc., of Washington, D.C., was also a ubiquitous MACOS critic. Several of the California creationists included MACOS in their speeches, although they did not fight the adoption of MACOS materials in the California curriculum once it was agreed that all references to evolution would be qualified by the phrase "many scientists believe that. . . . "[7]

A popular California journalist, John Steinbacher, traveled to communities lecturing about "the massive bulldozer operation to convert America's school system into a series of behavioral science classes for reshaping and restructuring children away from the Christian-Judaic tradition. . . . "[8] Syndicated columnist James Kilpatrick took a similar line in numerous editorials.[9]

The character of letters published in local papers or sent to congressmen also suggested links between apparently isolated textbook protests. The same Xeroxed articles were cited again and again. An unsigned information sheet about MACOS was widely circulated in the spring of 1975. It presented a list of objectionable points in the course and instructed citizens to write to their congressmen expressing these objections in their own words. Within a period of several weeks, congressmen received many similar letters objecting to the biased presentation of evolution, to the disturbing discussions of Netsilik behavior, and to the "atheistic philosophy" labeled variously "cultural relativism," "situational ethics," or "secular humanism."

As the letters arrived, a further question emerged. What was the federal government doing supporting such a course?

MACOS: A National Debate

"Is the Federal government supporting the subversion of American school children?"

The objections to the content of MACOS converged with the question of federal funding of public school curriculum to turn the textbook controversy into a national issue. Congressional interest in MACOS can be traced back to to 1971, when citizens groups in Phoenix, Arizona, complained about MACOS to their state senator John Conlan. Conlan, a Harvard Law School graduate and Fulbright scholar, was elected to the U.S. Congress in 1972 and remained interested in these concerns of his constituency. He hired as a staff aide, George Archibald, a writer for the *Arizona Republic*, who had focused on educational issues and served on the Arizona Board of Education's commission to develop guidelines for social studies texts. Archibald's goal is to "get schools out of the business of social engineering and indoctrination. . . . Schools exist for people, not for gurus."[10] He is convinced that the government should not support programs that have "any value orientation."

Congressional interest in MACOS was again apparent in 1973, when Republican representatives Marjorie Holt (Maryland) and John Ashford (Ohio) expressed their objections to "the usurpation by the educational system of what we used to consider parents' rights." Modern educators are competing with families, they claimed, and schools seek to "save children by bringing them into the arms of experts." They pointed to MACOS as one of several culprits.[11]

By late 1974, several Washington, D.C., organizations had begun questioning the federal government's role in developing and implementing MACOS. A Heritage Foundation report argued that MACOS taught a value system that was fundamentally deterministic, behavioristic, and relativistic. The course denied the existence of God and replaced traditional religious beliefs with concepts of Darwinian adaptation. School textbooks, claimed the report, must be either neutral, or present alternative values in a more equitable manner. The report cites letters from psychologists and therapists attributing cases of anxiety among children to the value conflicts provoked by MACOS. The government is blamed for MACOS, for without federal support, claims the report, "MACOS would fall on its own lack of merit."[12]

The Council for Basic Education also criticized the NSF for supporting a course that presented "cultural relativism and environmental determination" as a scientific explanation of the place of man in society. The children taking MACOS are shortchanged by its de-emphasis on skills and facts: "What is thrown out to make room for MACOS?"[13]

Leadership Action, Inc., mailed copies of selected "lurid" excerpts from MACOS to thousands of congressmen and state legislators. In April 1975, it presented a display of "100 dirty textbooks," in the Capitol, including mostly MACOS materials.[14] During the same month, a Washington, D.C., radio station placed an ad in the *Washington Post* for a brief NBC coverage of the dispute: "Tonight at 6:00: Horror Flicks—Is your 10-year-old watching X-rated films in school?"[15]

It was not, however, these moral objections to MACOS that influenced the largely urban and liberal congress. Rather, the anti-MACOS campaign appealed to the desire of many congressmen to control "unaccountable executive bureaucracies" such as the NSF; their resentment of scientists, who often tended to disdain congressional politics; and, above all, the concern with secrecy and confidentiality that followed the Watergate affair. The focus on MACOS during the NSF appropriations hearings in the spring of 1975 was more a reflection of post-Watergate morality than of Eskimo morality.

Conlan himself initiated the action against appropriations for MACOS on the grounds of its "abhorrent, repugnant, vulgar, morally sick content." It is "a godawful course" that was "almost always at variance with the beliefs and values of parents and local communities." It was "an assault on tradition," an attempt to "mold the children's social attitudes and beliefs and set them apart from the beliefs and moral values of their communities." It was, he claimed, the product of an "elite group of scholars" who want to "reform human nature from a behavioral program rather than along classic Judaeo-

Christian lines."[16]

Conlan brought these accusations to the House Committee on Science and Technology, seeking an amendment to the NSF appropriations bill that would deny the use of federal funds for the implementation or marketing of course curriculum programs unless the House Committee on Science and Technology and the Senate Committee on Labor and Public Welfare first approved the materials. To counter any such restriction, H. Guyford Stever, NSF director, wrote to the Committee Chairman, Olin Teague, announcing his intention of terminating all funds for MACOS and of placing a moratorium on the NSF implementation program pending the results of an internal review. Despite this letter, Teague initially supported Conlan's proposed amendment. Representatives Symington and Mosher, however, raised questions about the appropriateness of congressional censorship of specific programs. This censorship issue created interesting alignments within the Committee on Science and Technology. Its Democratic chairman (Teague) supported a proposal by its most conservative Republican (Conlan). Both were opposed by the ranking Republican (Mosher), whose position against the amendment was supported by most of the Committee's Democrats. In addition, a number of conservative members who basically agreed with Conlan on the issue of morality were reluctant to engage in censorship. The amendment thus lost within the Committee by a vote of 17–13.

Conlan changed his tactics when he brought the issue to the House floor on April 9, 1975. Avoiding the question of censorship, he focused on the federal role in implementing MACOS, on the accountability of federal agencies, and on the controversial nature of the social sciences. One must remember that this issue followed on the heels of Senator Proxmire's attack on NSF support of "crazy studies" in the social sciences; it followed the Watergate concern over public accountability and secrecy in federal bureaucracies; and it followed soon after dramatic and violent disruptions over textbook selection in Kanawha County, West Virginia. Thus a single elementary school course became the focus of national debate.

Several specific aspects of NSF's role upset the Congress. First was the marketing issue—the concern that the NSF used taxpayers' money to interfere with private enterprise. In fact, most publishers stayed out of this dispute, having themselves benefited from NSF programs, but a few had written to their congressmen to complain about "unfair competition" from a federal agency. A letter to the Committee on Science and Technology from Follett Publishers, for example, accused the Foundation of supporting a program that would fail to sell if it had to compete directly with the private sector. "Let the program die a natural death." Likewise, Lippincott wrote to Congressman Conlan, accusing the Foundation of unfair competition with the private sector when it adjusted the normal royalty arrangements in order to meet the high cost of the professional dissemination program.

A second and related complaint among congressional critics was NSF's use of resources "to set up a network of educator-lobbyists to control education throughout America." The flamboyant arguments of syndicated columnist James Kilpatrick, who attacked NSF's curriculum program as "an ominous echo of the Soviet Union's promulgation of official scientific theory," impressed a number of congressmen.[17] Resentment of the "elitism" of science reinforced concern that the NSF was naively promulgating the liberal values of the scientific community to a reluctant public. The market, it was claimed, was an adequate indicator of local preferences, and the reluctance of experienced publishers to market MACOS suggested what local communities really wanted. For the NSF to override market indicators by providing federal funds to implement MACOS was simply to support experts "who are trying to promulgate their own values." This was "an insidious invasion of local autonomy in education."[18]

MACOS defenders justified NSF support, arguing that the curriculum had been well-tested and acclaimed as a major educational advance. The issue, they argued, involved academic freedom. Can educators' freedom of access to the fruits of scholarship and educational research be limited by the views of a small group of people concerned with "anti-American motives"?[19] For Congress to restrict the NSF would be political censorship; to deny schools access to an existing curriculum would be an even greater restriction on local school board autonomy. What right had Congress to interfere?

"The Holy Bible would never pass muster under this kind of dema-goguery . . . in the Holy Bible there is murder, adultery, and there is bestiality in Little Red Riding Hood."[20]

A third issue in the congressional discussion of MACOS, having little to do with the course itself, was the desire to extend political control over "faceless, nameless executive agencies." Conlan com-plained about the difficulties he had encountered obtaining material from the NSF. "Somewhere, sometime, we are going to have to stop and make a stand as to whether the bureaucracy runs us or whether we represent the people and are accountable to them. I say this is a good place to start."[21] The uproar over "crazy" social science projects had reinforced the concern for accountability among con-gressmen, who often had to take the blame for decisions made by agencies they did not directly control. One congressman wrote to Stever about "those damn fool projects in the behavioral sciences," complaining that he was "sick and tired of responding to correspon-dence from citizens who are blaming Congress for some of the idiotic things done by a few unstable people in the executive branch."[22]

These issues converged in the congressional debates on NSF ap-propriations on April 9, 1975. After three hours of debate, the Con-lan amendment was defeated on the House floor by a vote of 215 (182 Democrats, 33 Republicans) to 196 (89 Democrats, 107 Repub-licans). While the House would not approve the kind of congression-al auditing of educational materials proposed by Conlan, it did pass an amendment requiring that all instructional materials, including teachers' manuals, be available to the parents of children engaged in NSF-supported programs.[23] Moreover, funds for MACOS were ter-minated and further support of science curriculum projects sus-pended, pending review of the entire NSF educational program. The congressional mood was clear when, on the same day, the House passed the notorious Bauman amendment—a much broader measure that would have involved Congress directly in the NSF grant applications review process, giving it veto power over *all* pro-posed grant awards. Despite the impracticality of congressional re-view of some 15,000 often highly technical proposals each year, Congress approved this amendment by a vote of 212–199. It was killed by a Senate Conference Committee before reaching a vote in the Senate, but neither NSF nor its congressional critics failed to realize its importance as a sign of political attitudes toward the NSF and its potential threat to the autonomy of science.[24]

These attitudes related to the broader political interests of several congressmen. Conlan intended to run for the Senate in Arizona where the Republican vote is heavily committed to Sam Steiger. Needing an appealing grassroots issue, he ran the opposition to MACOS as if it were a political campaign, collecting names from the thousands of anti-MACOS letters that he received and sending each person anti-MACOS materials. To Olin Teague, the MACOS dispute was also a significant political issue. At first he had supported Conlan, but once his Science and Technology Committee voted against "censorship of MACOS," he remained silent. However he did retain an intense interest in the issue. Teague was in a difficult position due to the declining importance of the space program so vital to the interests of his Texas constituents.

As chairman of the House Committee on Science and Technology, Teague must work in close liaison with the liberal scientific community, but his bedrock support comes from conservative Texans, who have had a substantial interest in textbook politics. Similarly, Representative James Symington from Missouri, also a member of the Committee on Science and Technology, risked criticism from his constituents. An article in the conservative journal *Human Events* accused Symington of "manipulating unsuspecting schoolchildren into the rejection of long established values," and suggested that his support of MACOS would affect his chances to succeed his father in the Senate.[25] The article was widely circulated in Missouri, where Symington was already in trouble for supporting abortion.

After the hearings in the spring of 1975, other consequences of the MACOS dispute became apparent. Sales of MACOS, of course, declined. The drop in sales during the second quarter of 1975 suggests the impact of congressional hearings (Table 10). EDC, the educational center responsible for MACOS and for much of the science curriculum development of the 1960s, continued to submit proposals to various government agencies. Every EDC proposal was turned down, and the center has turned to private sources of support for educational innovation.

During the summer of 1975, congressmen also attacked another NSF scientific program called Individualized Science Instruction System (ISIS). This course reflected NSF's changing philosophy of curriculum development, and its efforts to appeal to a wider student body. ISIS had originated at a conference organized by scientists from Florida State University in October 1971 to discuss how to make science more palatable to high school students at a time when so many of them were "turned off." The scientists proposed a program of eighty, two-to-three week minicourses or "modules" focusing around particular examples of science applied to contemporary problems. NSF funded the program at $3.3 million. By 1975, several of the modules were being tested in local schools. One of these, dealing with human reproduction, included a sophomoric presentation of human sexuality reminiscent of the Woody Allen film, *Everything You Always Wanted to Know About Sex but Were Afraid to Ask*. In testing this film, the developers, aware of potential opposition, included a letter warning teachers to evaluate its appropriateness in their community. Someone mailed a copy of this letter to Conlan, who tried to see the film. Unsuccessful, he complained that the film "was apparently so hot that it was unavailable for parental or congressional review."[26]

Accusations go beyond the question of sex education. Critics argue that ISIS presents even the physical sciences from a biased social perspective that reflects professional prejudices; environmental considerations, for example, dominate the presentation of materials in geology. According to Conlan's aide, this is a "Nader approach to education. . . . Facts should be taught and the student allowed to make up his own mind."[27]

The Tip of the Iceberg: MACOS, Accountability, and the NSF

By October 1975, the three committees that had been reviewing the NSF curriculum program had published their reports. NSF's internal review committee concluded that the Foundation could not avoid controversy at the expense of educational and scientific values, but that tighter review and evaluation procedures were in order. It made both procedural and policy recommendations. A "needs assessment" program should be initiated in order to establish priorities that would guide funding for curriculum development activities. Procedures must guarantee that programs are funded on the basis of broad solicitation for proposals and competitive review. The committee also recommended review mechanisms that would include participation by public representatives, but total responsibility for adoption of curriculum must rest with local or state authorities. In its policy recommendations, the NSF committee asked the National Science Board to develop a clear policy statement concerning science curriculum activities and the role of the Foundation in the implementation of potentially controversial materials. Finally, it recommended a study to develop new implementation approaches that would involve local authorities and allow NSF "to remain at arm's length from the process."[28]

A committee appointed by Congressman Olin Teague and chaired by T. M. Moudy, chancellor of Texas Christian University, also reported in October. While largely reflecting Teague's links to the space program and Texas, it included a spectrum of opinion ranging from a fundamentalist Texas housewife and several people in the Texas aerospace industry to Gerard Piel, publisher of *Scientific American*. The committee held hearings and conducted a mail inquiry to local school districts in fourteen states, receiving a largely positive response. The letters from local school superintendents and principals were enthusiastic, although some noted they had omitted sections of the course as "not in keeping with the mores of their community," while others observed that local pressures had caused the course to be dropped.

Some committee members criticized MACOS as "a venture in applied behavioral psychology without informed consent." They were distressed with the "presentation of evolution as a fact," with "cultural relativism," and with discussions that were "offensive to Judaeo-Christian values." One person wanted NSF out of the curriculum development business entirely; two others proposed that NSF restrict itself to funding only natural science and mathematics. The committee, however, with one formal dissent, recommended that the NSF continue to fund precollege science curriculum activities, advising only that it must avoid the appearance of "exercising such undue influence on local curriculum decisions." Thus, initiative for projects must come from local institutions. Moreover, "representative parents . . . innocent of professional or scholarly bias" should be appointed to curriculum review and evaluation groups so that they are involved in decisions that may affect widely observed customs or religious beliefs. The Moudy committee also warned the Foundation to caution teachers about their handling of cultural differences; teachers must honor the diverse value systems of the homes from which their pupils come. Finally, claimed the committee, the NSF cannot avoid responsibility for the quality and the content of the curriculum that it funds.[29]

The General Accounting Office (GAO) conducted a third review of NSF in response to accusations of improper financing. It cleared the Foundation, noting only a lack of clarity in its policies concerning royalties. It recommended strengthening NSF's review process, and then added a new wrinkle: children involved in new courses could be regarded as the objects of behavioral experiments, and yet the NSF had no procedures directed to their protection as human subjects.[30]

As the various committees examined MACOS and emphasized improved and more broadly participative review procedures, congressional critics extended their attacks on other NSF policies. In July, the House Science and Technology Committee held oversight hearings on the NSF peer review system. Conlan's frustration over the Foundation's refusal to identify MACOS reviewers led to accusations that the NSF had misrepresented the peer review evaluations. On September 17, Conlan announced his objections to NSF's funding of "low priority behavioral research and curriculum projects." There was, he claimed at an interview with two members of the National Science Board, no excuse for funding so much social research instead of other work that would help industry create jobs.

Two weeks later, Conlan and Senator Jesse Helms (R.-North Carolina) simultaneously introduced bills to the House and Senate asking for an amendment to the NSF Act that would require a peer review and grant management system that was "fair, open, and accountable to the scientific community and to Congress."[31] "MACOS is the tip of the iceberg . . . the beginning of what may be a wholesale reform of the massive NSF operation . . . the secret arbitrary system that does not preserve the best interests of U.S. science generally or the taxpayers who provide NSF's annual budget of about 750 million dollars." NSF's policies, claimed Helms, are "contrary to the American democratic spirit, the principles of justice and fairness. . . . At a time when the Congress is embracing openness as an antidote to Watergate, NSF stands firm for confidentiality."[32] Conlan and Helms called for disclosure of proposal reviews. When Stever responded that he would send edited versions of reviews to Congress, they compared his offer with Nixon's providing Congress with edited White House transcripts.

Pressure on the NSF increased as a GAO followup study of the peer review process in the Directorate of Science Education sustained Conlan's contention that the review of ISIS had been misrepresented. In response, NSF appointed a panel of seventy-three scholars, educators, publishers, and parents (including people who had been critical of the Foundation) to review nineteen of its precollege projects. Five of the projects were dropped, but ISIS survived—minus several of its modules, including the one on sex.

NSF also responded with an internal reorganization. Prior to July 1975, all scientific divisions were under the assistant director of research. In July, the biological and social sciences, the two most controversial areas, were placed under a separate directorate. This was done to provide additional senior management attention to these areas, but the move provoked a paranoid internal rumor that the Foundation was preparing to lop off both controversial areas. The NSF has also tightened its review procedures in the educational division. An "action review board" reviews all awards and program guidelines; outside reviews are more structured and formalized; an assessment of need is required prior to funding; and formal competitive procedures are being developed. A new position of assistant director has been created; the assistant director's job is to work with Congress. Meanwhile, the educational division has been asked to ponder all potentially controversial areas within its own programs; at one point the staff came up with an incredible list including the evolutionary implications that could be found in studies of comparative anatomy, the racial implications of the study of genetics, and all the manipulative possibilities inherent in the social sciences.

Things were looking up during the NSF appropriation hearings in the spring of 1976. The House Committee on Science and Technology recommended a $9 million increase in the science education budget to stem the downward trend during the preceding fifteen years, but the precollege curriculum and materials development program was reduced from $3.5 million to $1.4 million. No new project starts were authorized. Meanwhile, NSF continues to debate its policies for review and approval of educational programs. Within the Foundation there remains a strong preference to treat educational development proposals like research proposals, evaluating them through the peer review system, and avoiding all monitoring on the assumption that collegial pressures are sufficient to assure the quality and neutrality of the scholarship. NSF hopes thereby to avoid imposing a federal "mark of approval" that would only imply its further intervention in local decisions. The Foundation, claimed Stever in the Moudy Committee hearings, does not intend to force its curriculum on anyone. In the case of touchy value questions, he claimed, "the best practice is to pick a central group of value judgments and to use the peer review system to evaluate what these may be."[33] The role of the NSF is simply to provide a model curriculum developed by the most competent scholars, to be adopted as desired at the initiative of local users.

Yet members of the National Science Board, as well as external critics, have urged NSF to monitor and control the quality of its educational products prior to dissemination. This poses a difficult policy dilemma. If the Foundation finds it cannot approve a curriculum for dissemination, it will be subject to accusations of censorship. If it does sanction a curriculum, this is indeed a "mark of federal approval," and NSF will be exposed to even more accusations of imposing federal standards on the educational system. And to the extent that NSF becomes involved in monitoring its educational program, it may also be driven to do the same for any other program with potential policy implications.

Notes

1. This case study is a report of actual events in a specific school, but names are changed. The discussion is based on interviews with the school principal, teachers, and parents in the district, and on records that were kept of meetings. Unidentified quotations are from interviews.

2. There had been a Christmas carol dispute in town the year before. The school board had ruled it was not compulsory to sing Christmas carols, and many parents had objected.

3. This was evident in correspondence to Education Development Center and Curriculum Development Associates, and in evaluations as well in national awards.

4. EDC, *Community in Conflict over Curriculum Change* (Cambridge, Mass: EDC, 1972). This is a record of clipping and letters pertaining to the Phoenix dispute brought together by EDC.

5. Records are available from local newspapers wherever there were disputes (e.g., *The Burlington Free Press*, throughout November 1973) and from minutes of meetings held with school board officials.

6. The quotes that follow are from newspapers, records of disputes, interviews, and from minutes of meetings.

7. The state approval in California came about in part because of the fragmentation of the forces that would have opposed the adoption of the series. Legislative changes in the textbook selection process required consideration of racism, sexism, and other issues in textbooks; the board had to contend with many new interest groups, and the concern with evolution had been diffused. In addition, MACOS was already used in fifty-six schools in the state, which had adopted the series through local decisions.

8. Interview with Steinbacher reported in the *Green Mountain Gazette*, 7 November 1973. Steinbacher's philosophy is laid out in *The Conspirators: Men Against God* (California: Orange Tree Press, 1972).

9. These appear in local newspapers, for example, *Troy Times Record*, 3 April 1975; *Boston Globe*, 27 March 1975 and 2 April 1975.

10. George Archibald, personal interview.

11. Cited in *Review of the News*, 24 October 1973.

12. Susan Marshner, *Man: A Course of Study—Prototype for Federalized Textbooks?* (Washington, D. C.: The Heritage Foundation, July 1975).

13. Council for Basic Education, *Bulletin* 19 (May 1975): p. 9.

14. Leadership Action, *Special Textbook Report* (Washington, D. C.: Leadership Action Inc., n.d.).

15. *Washington Post*, 24 April 1975.

16. Congressman Conlan, during a House session authorizing Appropriations to NSF, *Congressional Record*, 9 April 1975, H2588.

17. Editorial in *Boston Globe*, 2 April 1975.

18. Congressman Conlan, *Congressional Record*, 9 April 1975, H2588.

19. Peter Dow, "Open Letter from EDC," 4 April 1975 (mimeographed).

20. Congressman Ottinger, *Congressional Record*, 9 April 1975, H2591.

21. Congressman Conlan, *Congressional Record*, 9 April 1975, H2603–4.

22. Reported in *Science*, 4 July 1975, p. 26.

23. *Congressional Record*, 9 April 1975, H2606. However, an amendment to make materials available to members of local communities (other than parents) in advance of implementation failed to pass on the grounds that it would be a vote of "no confidence" in the discretion of school boards.

24. *Congressional Record*, 121, no. 124, 30 July 1975.

25. J. F. Baldacchino, Jr., "MACOS: Pupils Brainwashed with Federal Funds," *Human Events*, 10 May 1975.

26. Note in *Human Events*, 9 August 1974.

27. George Archibald, personal interview.

28. National Science Foundation, *Pre-College Science Curriculum Activities of the NSF: Report* (Washington, D. C.: NSF, May 1975).

29. J. M. Moudy, chairman of Science Curriculum Implementation Review Group, *Report to the Committee on Science and Technology*, U.S. House of Representatives, October 1, 1975.

30. General Accounting Office, *Administration of the Science Education Project "MACOS"* Report No. MWD-76-26, 1975.

31. Bills, House of Representatives H.R. 98921, September 29, 1975. Senate S 17003, September 29, 1975.

32. *Congressional Record* 121, no. 144, 29 September 1975.

33. Hearings of the Science Curriculum Implementation Review Group (mimeographed).

IV SCIENCE AND THE RESISTANCE IDEOLOGY

Today's American ideologue is a middle-class man who objects to his dependence on science even when he accepts its norms. He is resentful of the superiority of the educated and antagonistic to knowledge. His ideology is characteristically not of the left but of the right. It . . . looks back to a more bucolic age of individuality and localism, in which parochial qualities of mind were precisely those most esteemed—to a simple democracy, in fact . . . it is the resistance ideology of all those who hitherto were the "stand" figures of our society in an earlier day; the models of once sober, industrious, and responsible citizens. —David Apter, *Ideology and Discontent*

8 Social Sources of Textbook Disputes

Recourse to religion is to the promise of science as recourse to the Roman courts was to the promise of Christianity. The older institution in either case was expected to pass away as a great promise was fulfilled. Neither promise was fulfilled and both institutions survived.[1]

In the very center of the most advanced technological societies there has emerged a resistance to science: an ideological resistance to the rationality and reductionism epitomized by science and a political resistance to its pervasive influence as a leading social institution.[2] While the origins of textbook disputes may lie in the regressive and often obscure demands of religious fundamentalists, this analysis suggests that they are manifestations of a much broader social phenomenon—widespread concern with the social implications of technology, and a hostility toward public institutions and expertise that pervades American society. Creationists and other textbook watchers are not numerous, but they are politically effective to the extent that they express social dissatisfactions that dispose people to seek alternative explanations of the human condition. Like anti-fluoridation disputes or protests against the siting of nuclear power plants or airports, textbook controversies reflect a resistance to impersonal, expertise-dominated bureaucracies that fail to respond to public priorities.[3] Furthermore, the textbook watchers, like many of the people preoccupied with mysticism, astrology, or various pop cosmologies and Eastern religions, question the image of science as an infallible source of truth.[4]

Resistance to science assumes different forms: some groups seek new consciousness through meditation or mysticism; others through a return to traditional values. But the combination of science and religion—as when creationists set theology in a context of scientific method, research monographs, and professional societies—is characteristic of many groups resistant to science: yogis use electroencephalographs to assist in meditation; *The Exorcist* shows doctors using sophisticated scientific medicine to exorcise spirits from a child possessed by voodoo; Scientism and UFO cults use scientific apparatus; and even occultists seek scientific validation for their beliefs.[5] The resistance to science is thus imbued with science. People are less concerned with its value as a useful activity than with its institutional and philosophical implications, which seem to deny man a sense of place and priority.

In order to understand the creationists and the attacks on science education programs, therefore, one must go beyond the pejorative labels of antiscience and irrationality to analyze the social and political tensions that sustain the resistance to science. These tensions are expressed in three themes that pervade many disputes over science and, in particular, the textbooks disputes described in this volume:
1. disillusion with science and technology as threats to traditional values;

2. resentment of the authority represented by scientific expertise as it is reflected in public school curriculum decisions;

3. defense of the pluralist and egalitarian values that appear threatened by modern science.

Disaffected with the scientific approach to nature as taught in the new science curriculum, both creationists and the anti-MACOS groups sought to incorporate traditional values into education, rejecting decisions by experts (the university scholars who created the curriculum), and applying an egalitarian ethos to educational decisions.

Disillusion with Science and Technology

It seems to me that man is not getting better, but is developing more diabolical ways of hurting his fellow man. Moreover because of his ineptness and his inability to see all the implications of his actions, he misuses many of his scientific accomplishments. He is hardly in a position to guide evolution.[6]

The textbook movement has been most active in southern California, Texas, and several western states that have experienced extraordinary population growth and economic fluctuations associated largely with science-based industry.[7] Indeed, many textbook watchers are engineers employed in the aerospace industry, people

who have personally experienced the discrepancy between technological expansion and the ability to deal with the social and economic
problems induced by rapid change. They are particularly distressed
with the uncertainties and disruptions of modern society, and they
associate a "decline in moral and religious values" with the dominance of scientific and secular perspectives. Hippies, campus re-
volts, venereal disease, drugs, and environmental problems, they
argue, all reflect "the liberalism of a scientific age."

Vernon Grose, the systems engineer who revised part of the California *Science Framework* to include creation theory in biology textbooks, associates evolution theory with

a campaign of secularization in a scientific-materialistic society—a
campaign to totally neutralize religious convictions, to destroy any
concept of absolute moral values, to deny any racial differences,
to mix all ethnic groups in cookbook proportions, and finally the
latest—the destruction of the distinction between male and
female.[8]

All this, he claims, is nonsense. Differences were created by design;
the source of our problems is the secularization of religious beliefs.
Similarly, Congressman Conlan argues that a philosophy that assumes all values and moral issues to be open and relative should be
blamed for the nation's most pressing social problems.[9]

Three aspects of the biology and social science curricula appear to
be especially distressing for their social implications: the biological
determinism that is implicit in the relationship drawn between animals and man, the implication that values are relative to situational
factors, and the denial of an omnipotent and omniscient force that
determines human behavior. Creationists argued that an emphasis
on the genetic similarities between man and animals is a socially dangerous concept that encourages animal-like behavior:

If man is an evolved animal, then the morals of the barnyard and
jungle are more natural . . . than the artificially imposed restrictions
of premarital chastity and marital fidelity. Instead of monogamy,
why not promiscuity and polygamy? . . . Self-preservation is the first
law of nature; only the fittest will survive. Be the cock-of-the-walk
and the king-of-the-mountain. Eat, drink and be merry, for life is
short and that's the end. So says evolution.[10]

One woman blamed the "streaking" craze on the theory of evolution. "If young people are taught they are animals long enough, they'll soon begin to act like them."[11]

Several assumptions that were only implicit in the teaching of evolution became explicit in MACOS, as it used examples of animal behavior to develop concepts about the nature of man. Objections were vociferous. Why should an explanation of man concentrate in such detail on animals? This was, the MACOS critics suspected, a pernicious attempt to spread a religion of "secular humanism."[12] Studies of how Netsilik values were influenced by a difficult environment reinforced their concern. To deny absolute standards by emphasizing adaptation to environmental pressures was, to the conservative critics of MACOS and to the creationists, socially irresponsible, morally destructive, and even harmful to family life. "Already we condone feticide and abortion and devalue human life, all in the name of biological principles."[13]

Unlike nineteenth-century fundamentalists, most modern-day creationists are willing to compromise by balancing the "pernicious influence" of modern science with a presentation of alternative theories. "Let the students take their choice." But the direct denial of absolute values in MACOS met the worst expectations of the textbook watchers. Their children were taught that in some societies senilicide and infanticide were functional, that their own values were only relative, based on specific environmental and situational factors in an indeterminate universe. These assumptions had profound implications for belief in the omnipotence of a God who determines human behavior. Picking up on contemporary public issues, they argued that "the denial of the existence of a designer serves as a license to destroy nature, fostering the self-centered behavior that is responsible for environmental problems." Recognizing man as God's steward would help man to understand and live in harmony with nature—to preserve given forms.[14]

Both creationists and anti-MACOS groups expressed their own sense of disillusion with science, but both also touched on popular anxieties about science and technology. However, these textbook watchers are not necessarily against technology: on the contrary, most of them earn their livelihood in technological industries.[15]

Probed for their opinions on currently controversial technologies, such as nuclear power or fluoridation, creationists welcomed technological growth as evidence of social progress. Yet their arguments captured widely held concerns about the social implications of modern science and technology, and especially about the role of authority and expertise in a democratic society.[16]

Challenges to Authority

An elite corps of unelected professional academics and their government friends run things in the schools.[17]

Schools exist for people, not for gurus.[18]

Modern science developed as a revolt against the authority of "sacred tradition"; yet it has become, for the science textbook critics, a symbol of an authoritarian ideology that suppresses such tradition. They express extraordinary resentment of "scientific dogmatism," of the "arrogance" and "absence of humility" among scientists. A sympathetic journalist writes of his great joy in "seeing science humbled," of seeing a blow to scientists' monopoly of truth."[19] In a speech to the California State Board of Education, a creationist observes, "After all, scientists put on their trousers in the morning one leg at a time, just like the rest of the world."[20] And a Jehovah's Witness writes about the "arrogant authoritarianism required by evolutionists to sustain what they cannot prove."[21] These concerns are not unique to textbook watchers. The resentment of professional arrogance is widespread; citizens at a public meeting of the American Nuclear Society complained of "the arrogance of scientists from the nuclear industry."[22] The condescension of medical professionals appeared to have influenced jurors at the Edelin abortion trial.[23] Thus, like the general concern with the impact of science and technology, this theme, too, reached a receptive audience.

The resentment of authority appeared in criticism of the academics who determine the content of school texts, and in the demands for local control of school curriculum. This was the basis of textbook revisions imposed by the California State Board of Education, revisions that guard against what is perceived as imposition of scientific authority beyond its appropriate limits. Particularly irritating to textbook watchers was the role of NSF, a "federal authority," in supporting the development of textbooks. It was this "federal takeover of education" that provoked the Willoughby lawsuit against NSF for funding BSCS materials, and it aroused congressional indignation in the MACOS dispute about

the insidious attempt to impose particular school courses . . . on local school districts, using the power and financial resources of the Federal government to set up a network of educator lobbyists to control education throughout America. . . . We Americans place a high value on local autonomy. Local school boards, reflecting the prevailing social norms of the community, should be the final arbiter of curriculum development.[24]

The desire to preserve parental control over textbooks motivated the two women who launched the creationist textbook movement in California. Increased parental participation in the selection of textbooks is the theme of the Gablers' organization in Texas. "Unless people take an active voice in assisting the authorized units of government in the process of selecting textbooks, the selection will continue to deteriorate."[25] The coordinator for the National Coalition for Children resents the "elitist groups of educators" that have usurped power from parents. Those opposing MACOS demanded that new curriculum be opened to public referendum, or at least review, a goal reflected in the congressional amendment to the 1975 NSF authorization act, requiring that all NSF-supported educational material be open to parental examination.[26] The Moudy Committee appointed by Congressman Teague also emphasized participation when it recommended that "representative parents" with no professional bias be involved in curriculum decisions that impinge upon widely respected customs and long-held religious beliefs.[27]

Similar feelings about "professional and scholarly bias" appeared in Kanawha County where the local people wanted to "get the government down to where they'll listen to us little old hillbillies." Education was a community issue, and powerful professionals were not to be trusted.[28] When a National Education Association panel investigated the issue, the Kanawha County Board of Education questioned the right of a Washington-based organization to judge the

mentality and intelligence of Kanawha County parents, especially
with regard to moral and spiritual values.[29]

The Supreme Court decision on pornography (that definitions of
obscenity are a matter for local determination) greatly stimulated
and sanctioned demands for local control. The textbook watchers,
concerned with "smut" in the schools as well as with science, had
watched this case with interest and delight, using it to back their
argument that, indeed, any program at all that impinged on local
values should likewise be judged within the community.

The public education system is one of the last grassroots institutions in America. School systems have traditionally been decentral-
ized, run by local school boards composed of elected citizens
(nonprofessionals). There has been, however, a gradual erosion of
local control through court decisions, reliance on nonlocal funds,
merging of school districts, and the general trend toward profes-
sionalism in education. School curriculum is guided more by national testing standards geared to college entrance requirements than to
local values. Those who retain the expectation of local control over
educational policy are threatened by the growing power and increasing professionalization of statewide curriculum committees
and departments of education. The federally funded programs created by academics who were isolated from local values and who were
attempting to develop a "teacher-proof" curriculum were a further
insult.

Centralization of curriculum decisions involves some assumptions
concerning public choice. Public school policy is binding on the individual. Once a curriculum is established, every student is involved.
Statewide policies are developed either on the assumption that the
knowledge in question is objective and indisputable, or that a given
policy is of broad social benefit, reflecting dominant values and
shared objectives regarding the purpose of education. The objection
to evolution as "dogma" suggests that, in science as well as other subjects, unresolvable value differences may preclude consensus.[30]

Insofar as local values influence curriculum decisions, they often
do so through watchdog citizens groups. Such groups have several
axes to grind: some claim that professionalization and standardiza-
tion of education favors only middle-class, college-bound students;

others are concerned with increased taxes; and still others have ideological or specific educational interests. Recently, the declining role of local control has been a source of tension more in matters of ethnic balance and busing than in curriculum and textbooks, but an important aspect of the textbook protest is that it parallels other developments. Thus, the protest appeals to diverse groups concerned with increasing government activity in education and more broadly with the lack of public discussion of issues vitally affecting community interests. Textbook critics have been quick to use the lessons learned from Watergate: "After the Watergate experience, this country is sensitized to those who appear to decline to open up an issue for full discussion. Today neither presidents nor scientists can expect to avoid an issue by virtue of the dignity of their office or profession."[31] In this context, any effort to challenge government or scientific authority, or to seek greater representation of community values can count on a wide base of support.[32]

The Ideology of Equal Time

Let us present as many theories as possible and give the child the right to choose the one that seems most logical to him. We are working to have students receive a fair shake.[33]

In January 1973, Henry Morris, president of the Institute for Creation Research, wrote to the director of the BSCS and challenged him to a public debate on the scientific aspects of the creation evolution controversy. He proposed the following topic: "RESOLVED that the special creation model of the history of the earth and its inhabitants is more effective in the correlation and prediction of scientific data than is the evolution model."[34] Morris proposed that newspapers, telecasters and the general public be invited, permitting large audiences "to hear both sides and decide for themselves." The outcome, essentially would be determined by audience applause. The issues, claimed the creationist, are free public choice, equality, fairness. Creationists argue that since the Biblical theory of origins is scientifically valid, it deserves "equal time." When there are two "equally valid hypotheses" it is only "fair" that students be exposed to both theories and be allowed to choose for themselves. "We are working to have students receive a fair shake."

Those who deny the right to equal time are accused by creationists of closed-mindedness ("those narrow men, like people standing in the sunlight who argue there is no evidence for the existence of the sun")[35] or exclusiveness (maintaining their "privileged position . . . not by scientific evidence, but by authoritarian proscription").[36] "A young professor could never admit he was a creationist without losing his chances for tenure," claimed a scientist at the Institute for Creation Research.

> In all the history of science, never has dogmatism had such a firm grip on science as it does today with reference to evolution theory. Evolutionists control our schools, the universities, and the means of publication. It would be almost as surprising to find an anti-evolutionist holding an important professorship at one of our major universities, as it would be to find a capitalist occupying a chair at Moscow University.[37]

The concept of equal time originated with the FCC Fairness Doctrine, confirming the responsibility of broadcasters to afford "reasonable opportunity for the presentation of contrasting viewpoints on controversial issues of public importance."[38] Its basis lay in the idea that power in a knowledge society is controlled by those who can communicate and that there is inequality in the power to communicate ideas just as there is inequality in economic bargaining power.[39] But the concept of equal time has been extraordinarily difficult to define and implement. With limited broadcasting resources, it has proved impossible to evaluate what issues are "of public importance" and who is entitled to present their viewpoints. In fact, the doctrine has tended to inhibit the presentation of controversial ideas. Nevertheless, equated with fairness and justice, and rooted in the democratic impulse, it has had enormous appeal in American society, especially in the almost "evangelically egalitarian educational system."[40]

The concept of pluralism allows that minority groups have a right to maintain cultural and religious traditions in the face of pressures for conformity. In the last decade, for example, we have seen textbook changes instituted by minority groups and by women, who see the presentation of history as biased. Since education is perceived as a condition for equality, such groups seek to have their interests more equally represented to students. Their demands formed a model for those who see bias in science curriculum—the struggle against cultural hegemony is paralleled by the struggle against scientific conformity, as textbook watchers insist that science education must respect beliefs that are outside the dominant scientific culture.[41]

Demands for equal representation of diverse interest groups in the selection of curricula are encouraged by the broadly conceived intentions of modern education. The goal of education, clearly expressed in the teacher materials of the new curricula, is to prepare students for their role as citizens in a democratic society. The BSCS stated that the purpose of improving biology education was to better train students to cope with the problems they will face as individuals, as parents, and as citizens. MACOS developers were also explicitly concerned with far more than communicating information, and justified the teaching of the social sciences as a means of developing the understanding of human nature necessary for citizenship. Unfortunately, they expressed this purpose in terms that were bound to provoke the ire of traditionalists.

It will not do to dream nostalgically of simpler times when children presumably grew up believing in the love of God, the virtue of hard work, the sanctity of the family, and the nobility of the Western historical tradition. . . . We must understand . . . what causes people to love and trust rather than to fear and hate each other, how patterns of mutual support can be fostered without destroying individual initiative and how we can learn to shape competing loyalties to common ends.[42]

The sense of fairness implied by demands for equal time buttressed the antiauthority appeal of the creationist movement. When William Willoughby sued the NSF for its support of textbooks teaching evolution, he reported receiving letters from nonreligious people sympathizing with his desire for equality. It is simply not fair, argued his correspondents, that the views of some groups are dismissed in this way. The creationists also convinced many people that they were speaking in the name of open-minded academic inquiry and intellectual honesty. "It is in the highest tradition of science to allow opposing viewpoints to be heard."[43]

Just as creationists appealed for the fairness of including Biblical explanations in biology books, MACOS critics wanted equal time for alternative world views in social science books. The Heritage Foundation Report, for example, demanded either a value-free social science *or* equal time for the presentation of alternative values.[44] James McKenna of the Heritage Foundation, legal counsel in several textbook protests, asserted

Under the banner of science, value systems are being marched into the schoolroom with a shameless disregard of the will of the polity. . . . The truth dawning on parents, be they creekers and rednecks from West Virginia or goldcoasters from Connecticut, is that education is a sectarian occupation. . . . The single most perfect example of this in our lifetime is the allegedly scientific disquisition into the roots of man. It has in its time gone by the name of Darwinism, evolutionism, or transformism. It is the most sectarian operation of the allegedly scientific community in the history of man.[45]

According to the Moudy Committee, MACOS is open to allegations that it is promoting "an evolutionary and relativistic humanism," because it "allows no hints of patriotism, theism, creationism, or other explicit values." The applications of science, especially social science, are "inseparable from people's beliefs, from their theology, from their morality." The committee concludes that input from "representative parents" should be included in every "value-laden" curriculum.[46]

These demands bring questions normally resolved by professional consensus directly into the political arena. From a professional perspective, science education is an enterprise best organized by experts, for it is intended to provide students with the best available information; what is taught or how it is taught is considered a technical matter. For those concerned with the values conveyed to their children in the schools, however, questions of personal beliefs, theology, and morality may enter into all educational decisions; choice of curriculum then becomes a matter for public participation. Scientific merit does not reduce political turmoil; on the contrary, the persistence of textbook conflicts reflects the continuing tension between professional assumptions about uniformity in the teaching of certain subjects and the expectations of pluralism in an educational system that is directed to doing far more than simply providing factual knowledge to students.

The tensions expressed in the science textbook disputes represent a reaction mostly from a politically conservative population. But criticism of the dominance of scientific values and the role of expertise, as well as demands for increased local participation, can be found among people of a wide spectrum of political ideologies. Indeed, similar tensions generated the "advocacy politics" that emerged during the late 1960s, as planners, health care and consumer advocates, and environmentalists mobilized around their diverse causes. All these groups have expressed disillusion with technology and expertise, and their slogans are "accountability," "lay participation," and "demystification of expertise." The following table presents similar statements from several groups and helps us place the demands of textbook watchers in a larger perspective. (See Table 11.)

The demands of textbook critics for increased participation and the representation of plural interests in science curriculum decisions leaves us with difficult questions: How can one determine if textbook material is antireligious or in some way biased? What kind of constraints or standards can be imposed to balance academic judgments with local and individual religious or personal concerns? How, in fact, do the personal beliefs and political values of free choice and fairness affect the communication and evaluation of science?

Table 11. Parallel Concerns Among Diverse Critics of Science and Technology

Environmentalists	Medical Critics	Science Textbook Critics
Disillusion		
"With technology's gifts to improve man's environment has come an awesome potential for destruction."	"Psychiatry and psychology are used as direct instruments of coercion against individuals. Under the guise of 'medical methods' people are pacified, punished, or incarcerated."	"Man is developing more diabolical ways of hurting his fellow man. Moreover, because of his ineptness and his inability to see all the implications of his actions, he misuses many of his scientific accomplishments."
Critiques of Expertise		
"The technologist's training can stand in his way. There is a growing awareness that civilized man has blindly followed the technologists into a mess."	"Professionals often regard themselves as more capable of making decisions than other people, even when their technical knowledge does not contribute to a particular decision. . . . Professionalism is not a guarantor of humane, quality services. Rather it is a codeword for a distinct political posture."	"Never has dogmatism had such a firm grip on sciences as it does today with reference to evolution theory. An elite corps of unelected professional academics and their government friends run things in the schools."
Demands for Increased Participation		
"It doesn't require special training to keep a broad perspective and to apply common sense. Thus, for every technically knowledgeable [person] there is a layman activist. . . . Any group which has interests at stake in the planning process should have those interests articulated. . . ."	"Medicine should be demystified. When possible, patients should be permitted to choose among alternative methods of treatment based upon their needs. Health care should be deprofessionalized. Health care skills should be transferred to worker and patient alike."	"Let us present as many theories as possible and give the child the right to choose the one that seems most logical to him. . . . Local school boards reflecting prevailing social norms of the community should be the final arbiter of curriculum development."

Sources: Editorials and the popular literature circulated by various Earth Day groups, the Health Policy Advisory Center, and science textbook critics

Notes

1. John A. Miles "Jacques Monod and the Cure of Souls,"*Zygon* 9 (March 1974): 41.

2. John Passmore talks of a revolt against science and emphasizes its powerful emotional force. John Passmore, "The Revolt Against Science," *Search* 3 (November 1972). See also Theodore Roszak, *Where the Wasteland Ends* (New York: Doubleday, 1972). There is now an extensive literature on current concerns with science. See, for example, J. Ellul, *The Technological Society* (New York: Alfred A. Knopf, 1956); V. Ferkiss, *Technological Man* (New York: Braziller, 1969).

3. See, for example, Dorothy Nelkin, *Nuclear Power and its Critics* (Ithaca, N.Y.: Cornell University Press, 1971), and *Jetport* (New Brunswick, N.J.: Transaction Books, 1974).

4. Christopher Evans, *Cults of Unreason* (New York: Farrar, Strauss and Giroux, 1973), discusses some of these cults and their relationships to science. The pop cosmologies of Velikovsky and Von Daniken have had enormous appeal (*The Chariots of the Gods* sold over 20 million copies) especially to people seeking scientific explanations that leave room for supernatural intervention. There has been little work on the social origins of participants in various cults. But in 1973, the National Opinion Research Center (NORC) ran a national survey asking 1,467 people if they had ever "felt very close to a powerful spiritual force." Of these, 35 percent had had what they described as a mystical experience. These respondents were mostly college-educated, Protestant, over 40, and male. Mysticism, the authors conclude, was not a property of a maniacal fringe; it suggests the limits of modernization and rationality among the population. W. McCready, "A Survey of Mystical Experience," *Listening*, 9 (Autumn 1974).

5. See M. Truzzi, "Towards a Sociology of the Occult'" in *Religious Movements in Contemporary America*, eds. I. Zaretsky and M. Leone (Princeton: Princeton University Press, 1974).

6. John Klotz, Letter to the Editor, *Christian Century*, 1 March 1967, p. 279.

7. Some tentative correlations could be attempted between the textbook controversy incidents and areas of rapid population change. Arizona, for example, with a 25 percent increase in population between 1970 and 1975, has been an active area for textbook disputes. Similarly, the Texas and Southern California communities experiencing strong fundamentalist influence have also been subject to major population shifts.

8. Vernon Grose, "Second Thoughts About Textbooks on Sexism," *Science and Scripture* 4 (January 1974): 14ff. With respect to the increasing concern about secularization, the editor of *Christian Century* suggests that this reflects a "protestant paranoia" developed as America has been transformed from Protestant domination to a much more diverse outlook. See discussion in Will Herberg's "Religion in a Secularized Society," in *Religion, Culture and Society*, ed. Louis Schneider (New York: Wiley, 1964), p. 596.

9. John Conlan, "The MACOS Controversy," *Social Education* 39 (October 1975): 391.

10. *Acts and Facts* (a publication of the Institute for Creation Research), April 1974.

11. *New York Times*, 10 March 1974, p. 49. "Streaking" was an obscure, shortlived craze in American universities during 1974 when students were "streaking" naked across public thoroughfares.

12. "Secular humanism" is a term loosely used by creationists and anti-MACOS groups to describe a world view that is human-centered and secular, emphasizing man's ability to achieve self-realization through the use of reason and scientific method and denying the importance of a spiritual and moral order. It implies that ethical standards should be determined by present human interests without reference to religion. On 12 May 1976, the House of Representatives passed an amendment to the National Defense Education Act (vote: 222–174) that "no preference be granted to the religion of secular humanism over the Judaic-Christian viewpoint" in government-supported curriculum.

13. Personal interviews by the author with creationists in San Diego, California.

14. The relationship between religion and environmental values has provoked much discussion. Some scholars attribute the environmental crisis to the Judaeo-Christian tradition that assumes that nature exists to serve man. They argue that Western science and technology is "cast in a matrix of Christian theology," that the creation story in Genesis justifies man's subjugation of nature. See Lynn White, Jr. "The Historical Roots of Our Ecological Crisis," *Science* 19 March 1967, pp. 1203 ff. The idea that Christian belief necessarily accepts the exploitation of nature, however, must be challenged by many counterexamples. The Southern Agrarians, while often fundamental Christians, saw the "Gospel of Progress" as an "unrelenting war on nature—well beyond reason," and advocated harmony with nature. See John Crowe Ransom, "Reconstructed but Unregenerated," in *I'll Take My Stand: The South and the Agrarian Tradition*, Twelve Southerners (New York: Harper & Row, 1930), p. 8. Others argue that the creation story implies responsibility and stewardship. See C. F. D. Moule, *Man and Nature in the New Testament*, (Philadelphia: Fortress Press, 1967). Moule claims that in Genesis the land belongs ultimately to God, and man is its trustee. The created order is valuable in itself—to be cared for by man.

15. The ambivalence among professionals working in high technology industries became apparent during the California Initiative on nuclear power, in which an organization called Creative Initiatives Foundation took an active role. This is a religious sect, and most of its members are upper middle-class professionals disaffected with the values of modern society. They seek new meaning in a life based on Biblical teaching and in efforts to change the value system of technology-based society. The three nuclear engineers who, with much publicity, left their jobs at General Electric to play an active role in the antinuclear campaign are members of this sect.

16. Disillusion with science and technology also reflects a loss of confidence in leadership and a general alienation and discontent. A Harris poll suggested that "general discontent factors" increased from 29 percent in 1966 to 55 percent in 1973. See discussion in A. Etzioni and C. Nunn, "The Public Appreciation of Science in Contemporary America," *Daedalus* (Summer 1974): 200–201.

17. John Conlan, "MACOS Controversy," p. 391.

18. Personal interview with George Archibald, congressional aide to Conlan.

19. Ron Kanigel, *San Francisco Examiner*, 16 September 1973.

20. Vernon Grose, *Statement to board of education*, in which he suggested revisions of the *Science Framework*, 1969, p. 4.

21. *Awake* (a Jehovah's Witness publication), 22 September 1974.

22. *Nuclear News*, May 1975, p. 83.

23. See articles by Barbara Culliton in *Science*, 31 January and 7 March 1975.

24. *Congressional Record*, 9 April 1975, H 2585-2587.

25. Educational Research Analysts, Florida (mimeographed brochure).

26. *Congressional Record*, 9 April 1975.

27. T. M. Moudy, Chairman of Science Curriculum Implementation Review Group, *Report to the Committee on Science and Technology*, U.S. House of Representatives, 1 October 1975.

28. A speaker at a demonstration in Kanawha, cited by Calvin Trillin, "U.S. Journal," *The New Yorker*, 30 September 1974, p. 121. While these protest groups challenge centralized professional authority, they are often strikingly authoritarian in their own ideas regarding appropriate educational style. For example, a prevalent objection to MACOS was its developmental, nonauthoritarian method of teaching, involving discussion and group participation. Its critics preferred a much more authoritarian approach to education.

29. National Education Association, *Kanawha County, West Virginia: A Textbook Study in Cultural Conflict* (Washington, D.C.: NEA, February 1975).

30. The problem of mounting religious commitment in a scientific society is a classic problem for minority sects. See Brian Wilson, "An Analysis of Sect Development," *American Sociological Review* 24 (February 1959): 3–15.

31. From a letter to the author in response to an article on textbook controversies.

32. Luther Gerlach and Virginia Hine, *People, Power, and Change* (Indianapolis: Bobbs-Merrill, 1970), attribute the increase in protest to "power deprivation." The reaction against professionals evident in our discussion suggests that these protests do tend to enhance the sense of social power among protesting groups. For a general discussion of the reaction to centralization and the appeal of local control, see Leonard Fein, *The Ecology of the Public Schools* (New York: Pegasus, 1971), and Alan Altshuler, *Community Control* (New York: Pegasus, 1970).

33. Personal interview with Henry Morris.

34. Letters from Henry Morris to William Mayer, 1 January 1973, through 15 April 1973. See *Acts and Facts*, June 1973.

35. George How, statement at hearing before the California State Board of Education, 9 November 1972.

36. Letter from Henry Morris to William Mayer, 1 January 1973.

37. Duane Gish, "A Challenge to neo-Darwinism," *The American Biology Teacher* 32 (February 1973): 495.

38. *Federal Register*, 10406, 1 July 1964.

39. Jerome Barron, "Access to the Press—A New First Amendment Right," *Harvard Law Review* 80 (1967): 1647. Proponents of the fairness doctrine claim that the First Amendment was written without the knowledge that communication could be dominated by a few private interests capable of practicing their own form of censorship over free expression.

40. This term was used to describe some sources of anti-intellectualism by Richard Hofstader, *Anti-Intellectualism in American Life* (New York: Vintage Books, 1962), p. 23.

41. That the teaching of biology poses a distinct problem for minority religious groups whose belief systems are incompatible with science has been recognized in court rulings exempting Amish children from school requirements. These decisions recognized that "the values of parental direction of the religious upbringing and education of their children in their early and formative years have a high place in our society." These can outweigh even strong state interest in universal education. See *Wisconsin* v. *Yoder*, U.S. 205, 213 (1972) discussed in Frederick LeClerq, "The Monkey Laws and the Public Schools: A Second Consumption?" *Vanderbilt Law Review* 27 (March 1974): 242.

42. Peter Dow, "MACOS: The Study of Human Behavior as One Road to Survival," *Phi Delta Kappan* 57 (October 1975): 81.

43. Letter from D. E. Martz to *Science* 9 March 1973, p. 953. A series of similar letters were printed in *Science* with respect to the creationist conflict. Similar perceptions are suggested by a poll of 1,200 teachers participating in the western area National Science Teachers Association convention in November 1972. Only 147 teachers responded (30 percent from biology, 21 percent from chemistry, 26 percent from general science, and 18 percent from physics) but the response suggests a perception of science that requires "fairness" in the presentation of scientific material. Fifty-seven percent agreed that alternative theories should be taught in public schools and 39 percent agreed that it was acceptable to present creation theory in science classes. Fifty-three percent disagreed with the statement that evolution theory and creation theory are mutually exclusive and that therefore only evolution theory should be included in science curriculum.

44. Susan Marshner, *Man: A Course of Study—Prototype for Federalized Textbooks?* (Washington, D.C.: The Heritage Foundation, July 1975).

45. James T. McKenna, "*Serrano* v. *Priest*: Where Have You Led Us?" *Imprimis* 3 March 1975, p. 13.

46. T. M. Moudy, *Report to Committee on Science and Technology*, U.S. House.

9 Science and Personal Beliefs

We resent the widespread philosophical prejudice that arises from the *a priori*, non-scientific assumption that there can be no divine intervention with the working of the universe and that the scientists' method is the only way to truth.[1]

Religious fundamentalists of the 1930s wanted to prevent the teaching of evolution theory because they saw it as an explicit threat to theological world views and traditional values. They succeeded in intimidating publishers so that evolution was virtually ignored in public school textbooks for thirty years. By the 1970s, renewed demands of fundamentalists, now called "creationists," had little influence on the firmly rooted discipline of biology, although the disputes may have reinforced existing reluctance to approach the subject in many classrooms. As one biologist pointed out, Darwin is by now quite immune to overthrow. On the other hand, opposition to the far more tenuous concepts in the social sciences was more successful, its influence extending well beyond the particular public school course to questions of research and educational policy. Concepts in social science appeared in the seventies much like evolution in the thirties—a direct and explicit threat to personal morality and religious belief.

Images of Science

The desire to extend personal beliefs and democratic principles to science that characterizes these textbook disputes suggests a public perception of science quite at odds with the perceptions of those directly involved in its practice. Scientists were amazed at the idea that questions of fairness in the representation of beliefs should determine the substance of scientific education: Can quacks be entitled to equal time? Should Christian Scientism appear in health books, the stork theory in books on reproduction, and astrological lore in expositions of astronomy?

Concepts of pluralism, of equity, of "fairness," of wide-open participatory democracy as practiced in a political context, appear to scientists to be incongruous with scientific education. Science is based on the assumption that nature is comprehensible by objective observation. Decisions are based on the existence of an organized body of knowledge, and on an intricate network of procedures accepted by a community of scientists who share values concerning appropriate behavior and standards of acceptable truth. These values are founded on a view of science as an autonomous system distinct from its political, personal, or social context. Robert Merton, whose work has laid the foundation of the sociology of science, described these values as universalism (claims of truth are subject to impersonal criteria); communism (the findings of science belong to the community of science); disinterestedness (claims are based on the testable character of science): and organized skepticism (methodological and institutional mandates require suspending temporary judgments until beliefs are tested in terms of empirical and logical criteria).[2]

In this context, scientific theories are taught because they are accepted by the scientific community as an objective explanation of reality. It is collegial acceptance that validates one theory and rejects another; acceptance or debate by those outside the community is totally irrelevant.[3] Moreover, while science is an open system in terms of social criteria, scientific recognition depends on achievement and rigorous evaluation; indeed, the internal standards of performance in science may run counter to egalitarian notions.[4]

The textbook controversies lead us to wonder about the public understanding of science. What do people know about science, about its process and its methods? Is public support of research based on any rational understanding of the scientific endeavor? How is scientific knowledge received by the public that supports it?

Years ago, Oscar Handlin suggested that science was hardly assimilated even in the cultures of advanced industrial societies. Those who voted federal funds for its support had no understanding of its character, for science had advanced beyond common sense and empirical reality.

Paradoxically the bubbling retort, the sparkling wires and the mysterious dials are often regarded as a source of a grave threat . . . the machine which was a product of science was also magic, understandable only in terms of what it did, not of how it worked. Hence the lack of comprehension or of control; hence also the mixture of dread and anticipation.[5]

Acceptance of the authority of scientific judgment continues to coexist with mistrust and fear. The romantic view of the scientist as "a modern magician, a miracle man who can do incredible things" is matched by negative images.

Dr. Faustus, Dr. Frankenstein, Dr. Moreau, Dr. Jekyll, Dr. Cyclops, Dr. Caligari, Dr. Strangelove . . . In these images of our popular culture resides a legitimate public fear of the scientist's stripped-down, depersonalized conception of knowledge—a fear that our scientists will go on being titans who create monsters.[6]

Ambivalence persists despite the post-Sputnik interest in science. In 1959, a public opinion survey reported that 83 percent of the American population thought that we are better off because of scientific contributions to health and to a higher standard of living. But 47 percent of the population also thought that science makes our way of life change too fast, and 40 percent feared that the growth of science would bring increased centralized control. Scientists were perceived as hard-working and brilliant, but "odd and peculiar" by 40 percent of the respondents, and the survey concluded that while there is a great deal of respect for science, there remains considerable mistrust and apprehension.[7] Recent surveys suggest even more pronounced ambivalence. In 1972, a Harris poll found that 76 percent of Americans worried about excessive concentration on science, and associated this with "neglect of human problems"; 72 percent believed that science was making us too dependent. Yet 89 percent of the respondents saw scientific progress as necessary for a high standard of living.[8] The British journal, *New Scientist*, surveyed a population of college students and found the scientist still stereotyped as remote and withdrawn, "a white-coated man in spectacles working in a laboratory." Those critical of science felt that the scientists' intellectual curiosity would usually triumph over moral responsibility, that the values of research are in direct conflict with human values.[9]

Ambivalent attitudes are often matched by confused comprehension. A series of surveys by the National Assessment of Educational Progress indicated that facts were far better assimilated and recalled than was understanding of scientific process; few respondents showed more than limited comprehension of the methods of scientific inquiry or the differences between facts, theories, and hypotheses. (See Appendix 2.) This is also apparent among those creationists who use the language of science but seem to understand little about its methods and underlying assumptions. Indeed, the textbook disputes illustrate two common notions about science that bear on its acceptance: (1) science can be defined as a collection of facts, and (2) it can be evaluated in terms of its influence and implications.

The belief persists that "value-free" truths can be derived from an accumulation of evidence. In this light, concepts in biology and the social sciences are especially prone to public criticism, for even theories that are well accepted within the scientific community may be based on untestable assumptions and indirect evidence. It is difficult to grasp how complex and indirect evidence may constitute support for certain scientific theories that cannot be verified by direct observation. Indeed, neither Darwin's contemporaries nor modern creationists understood that useful theories in science need not have definitive support if they have powerful predictive capacity. For those who fail to understand the complex and subtle relationships between fact and theory, acceptance of science may remain as fully a matter of faith as would commitment to supernatural explanations. And those with strongly entrenched religious beliefs or a sense of tradition that contradicts scientific explanations will reject these explanations and sometimes seek representation of their own views in the educational system.

Public acceptance of science is also related to the evaluation of its influence or potential implications. People tend to seek, in their beliefs about nature, the values that will guide their behavior.[10] It is efficacy more than evidence that affects the credibility of science, and this is encouraged by scientists, who sometimes claim excessive territory for the concepts and tools of their disciplines. This tendency is especially evident in recent efforts to extend concepts in the biological sciences to generalizations about man and society.[11] Such aggressiveness may be a source of success within science, but it may also encourage unrealistic public expectations and misconceptions.[12] Moreover, it makes it increasingly difficult to maintain the conventional distinction between science and its ideological and social content.

Extensive claims concerning the usefulness of science may also lead to the belief that any systematic practice that is in some way conceived to "work," such as faith healing, patent medicine, or transcendental meditation, may be viewed as scientific. This is widely manipulated by those interested in selling almost any product or service; useless products and regimens are justified as being based on scientific knowledge ("It's a medical fact").

Problems in the Communication of Science

The persistence of these images of science derives in part from problems in communication. Historically and methodologically, much of science developed in opposition to the dogmatism of religion, and most scientists understand their own work as approximate, conditional, and open to critical scrutiny. This is in striking contrast to the frequent public representation of science as authoritative, exact, and definitive. Science, claimed the original version of a California textbook, "is the total knowledge of facts and principles that govern our lives. . . . " (See Table 8.) The organized skepticism toward scientific findings that is tacitly understood by those who practice science contrasts sharply with its public image of dogmatism.

Perhaps the most difficult concept to convey to those who are not scientists is the delicate balance between certainty and doubt that is so essential to the scientific spirit. Textbooks especially tend to convey a message of certainty, for in the process of simplification, findings may become explanations, explanations may become axioms, and tentative judgments may become definitive conclusions.[13] Preoccupied with communicating a scientific view of nature, textbooks often neglect to convey concepts of critical inquiry. The new science curricula have tried to compensate for this tendency by emphasizing inquiry-oriented instruction, and the concepts and methods that characterize "real" scientific research. But the image of science conveyed in textbooks seldom includes analyses of the organization of research, the personal motivations of scientists, or the relationship of science to cultural and social attitudes.

Science is a cultural process and its theories are developed in the context of contemporary values.[14] Yet the individualistic tradition in science leads its practitioners to minimize the importance of the social and cultural processes involved. One historian of science has suggested, only partly tongue-in-cheek, that "the history of science be rated X," for a proper study of the historical development of scientific concepts and their underlying nonscientific values and assumptions would do violence to the professional ideals and public image of science.[15] Scientists themselves seldom speculate about the assumptions that underlie their work, and would indeed be paralyzed if they constantly had to question hypotheses that lie deeply embedded in the structure of their disciplines. When such attitudes are carried over to textbooks in the effort to convey the findings of contemporary research, problems of dogmatism are inevitably perpetuated. For the concepts and methods accepted by scientists and introduced as self-evident are by no means obvious to laymen.[16]

In a number of ways, then, the scientists who developed the new curriculum were as isolated from the mainstream of American values as the people in Kanawha, West Virginia. Insensitive to the social context in which their ideas were to be taught, they were hampered in their efforts to communicate science. Failing to grasp the differences between the structured, meritocratic processes within scientific disciplines and the more egalitarian, pluralistic processes outside science, they were unable to deal effectively with the political conflict. Biologists preferred to deny the conflict between science and religious values (e.g., by bringing clergymen to public hearings to argue the compatibility of religion and science), or they dismissed criticism of the teaching of biology and social science as evidence of ignorance or of educational failure.

Many scientists responded to criticism with their own kind of fundamentalism, emphasizing the neutrality and apolitical character of science and the weight of evidence that supported their authority. They expected a kind of literalism and realism when dealing with the religious claims of their critics, apparently forgetting that science itself is approximate and metaphoric. This literalism on the part of scientists when confronting opposition nearly matched the attitudes of fundamentalists, and it hindered both scientific and political communication.[17]

Underlying much of the scientists' defensiveness was a frankly political concern about external control over the definition of science as it would be taught in the schools. "The State Board's repudiation of its own committee in favor of a lay opinion from the audience should ultimately become a classic example in textbooks on school administration of how *not* to proceed with the development of standards," claimed an evolutionist.[18] "Why are comments related to science made by high-priced technicians such as medical doctors and by persons in related fields of technology more readily acceptable as statements of science than those made by scientists themselves?" complained another.[19] Scientists' response to external pressure in the 1970s was thus no different from their response in the 1930s, when they feared that "What is taught as science would be determined by . . . shopgirls and farm hands, ignorant alike of science." (See Chapter 2.)

Scientists are convinced of the rationality and merit of their methods and constantly dismayed by the popularity of nonscientific approaches to nature. In 1975, for example, hundreds of scientists signed a statement criticizing astrologers. They were puzzled that "so many people are prone to swallow beliefs without sufficient evidence," and concerned that "generations of students are coming out without any idea that you have to have evidence for your beliefs."[20] The persistence of creationism, the MACOS affair, and other textbook protests are reminders that beliefs need no evidence; that, indeed, people are most reluctant to surrender their personal convictions to a scientific world view.

To those whose personal beliefs are challenged, the social and moral implications that can be drawn from a scientific theory and its threats to the idea of absolute ethical values clearly may assume far greater importance than any details of scientific verification. Indeed, increased technical information is unlikely to change well-rooted beliefs, for selective factors operate to guide the interpretation of evidence, especially when the nature of such evidence is poorly understood.[21] Creationists, as we have seen, avoid, debunk, or disregard information that would repudiate their preconceptions, preferring to deny evidence rather than to discard their beliefs. A great deal of social reinforcement helps them maintain their views in the face of repeated frustration, and opposition only strengthens their religious convictions.[22]

The recurrence of textbook disputes suggests that the truce between science and religion, based on the assumption that they deal with separate domains, may be a convenient but unrealistic myth. Religion as well as science purports to be a picture of reality, a means through which people render their lives and the world around them intelligible. The heart of the religious perspective, argues anthropologist Clifford Geertz, is "not the theory that beyond the visible world there lies an invisible one; . . . not even the more diffident opinion that there are things in heaven and earth undreampt of in our philosophies. Rather it is the conviction that the values one holds are grounded in an inherent structure of reality."[23] It is clear that for many people, science, often unrelated to their experience, does not serve as a satisfactory explanation of reality on which to base their values. Failing to find a sense of personal integration from scientific beliefs, they seek alternative explanations.

Faith in science persists when it satisfies a social need. If science loses credibility ("planet earth is in trouble," the creationists claim), people will grope for more fulfilling constructs. Science threatens the plausibility of nonrational beliefs, but it has not removed the uncertainties that seem to call for such beliefs. The revival of fundamentalism fills a social void for its adherents. By using representations that are well adapted to the twentieth century, by claiming scientific respectability, or by arguing that science is as value-laden as other explanations, modern textbook watchers offer intellectual plausibility as well as salvation, and the authority of science as well as the certainty of scripture. Poorly understanding the process of science, they seek to resolve the old warfare between religion and science through popular decision. Democratic values such as freedom of choice, equality, and fairness become criteria for judging the merits of science. Textbook critics have thus managed to fuse three venerated traditions of American culture—science, religion, and populist democracy.

Notes

1. Letter from E. G. Lucas to *Science* 9 March 1973, p. 953.

2. Robert Merton, "Science and Technology in a Democratic Order," *Journal of Legal and Political Sociology* I (1942): 115–126. See also Norman Storer, *The Social System of Science* (New York: Holt, Rinehart and Winston, 1966).

3. See discussion of the relationship between evidence and theory in Harvey Brooks, "Scientific Concepts and Cultural Change," *Daedalus* (1964); S. B. Barnes, "On the Reception of Scientific Beliefs," in *Sociology of Science*, ed. S. Barnes (Harmondsworth: Penguin Books, 1972), p. 287.

4. See discussion in François Hettman, *Society and the Assessment of Technology* (Paris: OECD, 1973), p. 40.

5. Oscar Handlin, "Ambivalence in a Popular Response to Science," in *Sociology of Science*, ed. S. Barnes, p. 253, 267.

6. Theodore Roszak, "Science Knowledge and Gnosis," *Daedalus* (Summer 1974): 31.

7. Stephen Whittle, "Public Opinion about Science and Scientists," *Public Opinion Quarterly* 23 (1959–60): 383–388.

8. Harris Survey news release, 17 February 1972. For a review of various surveys, see Amitai Etzioni and Clyde Nunn, "The Public Appreciation of Science in Contemporary America," *Daedalus* (Summer 1974): 191–205. See also Todd R. Laporte and Daniel Metlay, "They Watch and Wonder: The Public's Attitudes Towards Technology," University of California, Berkeley, Working Paper #6, 1972; and R. Funkhouser and N. Maccoby, *Communicating Science to Non-Scientists* (Palo Alto: Institute for Communications Research, Stanford University, 1970).

9. Philip Hills and Michael Shallis, "Scientists and Their Image," *New Scientist*, 28 August 1975. This image is also conveyed in television programs, especially those on medical scientists whose scientific leanings invariably detract from their more "human" clinical involvements.

10. See discussion in Clifford Geertz, *Islam Observed* (New Haven: Yale University Press, 1968), and *Interpretation of Culture* (New York: Basic Books, 1973). Note that scientists as well as laymen often assume a congruence between scientific knowledge and values. This is most evident in ecology and population studies where it is assumed that increased knowledge will change values and behavior. (John A. Miles, personal communication, suggested this argument.)

11. R. M. Young, "Evolutionary Biology and Ideology: Then and Now," *Science Studies* I (1971): 177–206, discusses these tendencies to extend biology, and, in particular, the implications of the work of Ardrey and Lorenz.

12. The physicist Victor Weisskopf has discussed some of the implications of over-extending the scientific approach to deal with human experience in an address to the American Academy of Arts and Sciences, *Bulletin* 27 (March 1975).

13. James Raths, "The Emperor's Clothes Phenomenon in Science Education," *Journal of Research in Science Teaching* 10 (1973): 211. For problems in the communication of science, see also Philippe Roqueplo, *Le Partage du Savior* (Paris: Editions du Seuil, 1974). For reviews of the effectiveness of the new science curriculum, see A. L. Balzer et al., "A review of research on teacher behavior relating to science education," ERIC Information Analysis Center, Ohio State University, December 1973, pp. 87–93; D. Novak, "A Summary of Research and Science Education"; ERIC, December 1973, pp. 32–51. These studies also note the continuing emphasis on skills of recall in the science classroom despite the "inquiry-oriented" approach in the new curriculum.

14. This point is well developed by a study examining the history of the study of genetics in relation to widely held beliefs during different periods. See W. Provine, "Genetics and the Biology of Race Crossing," *Science*, 23 November 1973, pp. 790–798.

15. Stephen G. Brush, "Should the History of Science Be Rated X," *Science*, 22 March 1974, pp. 164ff.

16. According to one analysis of the new science curriculum, the new approach merely replaced the authority of facts with the authority of method, replacing "low cognitive-level dogmatism with high cognitive-level dogmatism." D. L. Gardner, "Structure of Knowledge Theory and Science Education," *Educational Philosophy and Theory* 4 (October 1972): 44.

17. Donald T. Campbell, "On the Conflicts Between Biological and Social Evolution and Between Psychology and Moral Tradition," *American Psychologist* 30 (December 1975): 1120. Campbell notes that "scientists hold up for religious discourse the requirement for a direct realism, a literal veridicality, even though they may recognize that this is impossible for science itself."

18. William Mayer, "The Nineteenth Century Revisited," BSCS Newsletter, November 1972.

19. David Ost, "Statement," *American Biology Teacher* 34 (October 1972): 413–414.

20. Statement by Paul Kurtz, editor of *The Humanist*, in justifying the statement by 186 scientists calling astrologers charlatans who have no rational basis for their beliefs. See *The Humanist*, October/November 1975.

21. Leon Festinger, *A Theory of Cognitive Dissonance* (Evanston, Ill.: Row Peterson, 1957).

22. Peter Berger, *Sacred Canopy* (New York: Doubleday, 1967), argues that a prerequisite for belief is social support—that definitions of social reality are real insofar as they are confirmed by day-to-day interaction.

23. Clifford Geertz, *Islam Observed*. The contemporary quest for "something to believe in" was certainly related to the character of the 1976 presidential campaign. John A. Miles, personal communication.

APPENDIXES

Appendix 1

NSF Precollege Science Curriculum Project Grants in Thousands of Dollars
(1957–1975)

Curriculum Project	Grantee	Develop-ment	Implemen-tation	Publisher, Date of Publishing Agreement
PSSC	EDC (MIT)	5,300	6,800	D. C. Heath (1959)
SMSG	Stanford U. (Yale)	14,400	2,260	None: Random House had distribution arrangement
CBA	Earlham College (Wesleyan & Reed)	1,200	2,300	McGraw-Hill (1962)
BSCS	U. Colorado (AIBS)	10,400	9,400	Houghton Mifflin; Rand McNally; Harcourt, Brace (1963–1964)
CHEM Study	U. California, Berkeley	2,600	4,600	None: W. H. Freeman, distributor
Elementary School Science Project	U. California, Berkeley	700	6,000	None: materials sold at cost
Elementary School Science Project (ESSP)	U. Illinois	600	6,000	Harper & Row (1968)
TV Program for Mathematics Teachers	Minnesota Academy of Sciences	200	—	—
Syracuse Webster Mathematics Project (Madison Math)	Webster College	1,100	2,500	—

Appendix 1. (continued)

Curriculum Project	Grantee	Develop-ment	Implemen-tation	Publisher, Date of Publishing Agreement
Elementary Science Study (ESS)	EDC	7,600	4,100	McGraw-Hill (1969)
Anthropology Curriculum Study Project	American Anthropological Association	1,400	700	Macmillan (1966)
Science—A Process Approach (SAPA)	AAAS	2,300	4,900	Xerox (Ginn) (1967)
Univ. of Illinois Committee on School Mathematics (UICSM)	U. Illinois	4,900	5,800	Macmillan; Harper & Row (1968)
MINNEMAST	U. Minnesota	5,000	700	W. B. Saunders (1969)
Science Curriculum Improvement Study (SCIS)	U. California (U. Maryland)	4,300	6,700	Rand McNally (1970)
Earth Science Curriculum Project (ESCP)	American Geological Institute	3,500	700	Houghton Mifflin (1967)
School Science Curriculum Project	U. Illinois	970	—	No commercial distribution
MACOS	EDC	4,800	2,200	CDA (1970)
Elementary School Science Improvement Project	Utah State	100	—	No commercial distribution

Appendix 1. (continued)

Curriculum Project	Grantee	Develop-ment	Implemen-tation	Publisher, Date of Publishing Agreement
Secondary School Science Project	Rutgers (Princeton campus)	1,200	650	McGraw-Hill (1966)
Introductory Physical Science (IPS & PSII)	EDC	1,400	5,000	Prentice-Hall (1965)
Films for In-Service Education of Teachers of Elementary School Mathematics	National Council of Teachers of Mathematics	280	—	Distributed by United World Films, Inc.
Quantitative Approach in Elementary School Science	SUNY, Stony Brook	21	11	No commercial distribution
High School Course Modern Coordinate Geometry	Wesleyan U.	137	—	Houghton Mifflin (1969)
High School Geography Project	Association of American Geographers	2,300	1,900	Macmillan (1968)
Sociological Resources for the Social Studies (SRSS)	American Socio-logical Association	2,500	1,800	Allyn & Bacon, Inc. (1968)
Engineering Concepts Curriculum Project (ECCP)	SUNY, Stony Brook (Commission on Engineering Ed., Polytechnic Inst. of Brooklyn)	2,000	3,000	McGraw-Hill (1971)

Appendix 1. (continued)

Curriculum Project	Grantee	Development	Implementation	Publisher, Date of Publishing Agreement
Elementary Mathematics Project	EDC	1,600	120	Under negotiation
Harvard Project Physics Course (PPC)	Harvard U.	900	4,700	Holt, Rinehart & Winston
Second Course in Physical Science (PS II)	Newton College of the Sacred Heart (EDC)	(See IPS-EDC)		Prentice-Hall (1970)
Portland Inter-disciplinary Science Project	Portland State U.	144	3	—
ISCS	Florida State U.	1,500	5,000	General Learning Corp. (1972)
Secondary School Mathematics Curriculum Improvement Study (SSMCIS)	Teachers' College, Columbia U.	700	190	Columbia Teacher's College Press (1974)
Computer-Based Self Instructional Course for Supplementary Training of Secondary School Teachers of Physics	U. California, Berkeley	139	—	—

Appendix 1. (continued)

Curriculum Project	Grantee	Develop-ment	Implemen-tation	Publisher, Date of Publishing Agreement
Environ-mental Studies for Urban Youth (ES)	Evergreen State College	900	800	Addison Wesley
Comparing Political Experiences (CPE)	American Polit-ical Science Association	1,300	57	Course under development
Improvement Project in Mathematics for Subcultural Groups	S. W. Education Development Lab.	400		No commerical distribution
Biomedical Inter-disciplinary Curriculum Project (BICP)	California committee on Regional Medical Program (U. Cal., Davis) (U. Cal., Berkeley)	1,850	160	No commercial distribution
Demonstra-tion and Experimen-tation in Computer Training	Dartmouth College	330		Prentice Hall (1971)
Boston University Mathematics Program	Boston U.	294		Under negotiation
Development of Computer Simulation Material	SUNY, Stony Brook (Polytechnic Inst. of Brooklyn)	470	150	Digital Equipment Corp. (1974)

Appendix 1. (continued)

Curriculum Project	Grantee	Develop-ment	Implemen-tation	Publisher, Date of Publishing Agreement
Development of Teacher Training Materials in Mathematics	State College of Iowa	24		Sold at cost by grantee
Experimental Teaching of Mathematics in Elementary School	Stanford U.	2,700		Academic Press (1971)
Exploring Human Nature (EHN)	EDC	2,500	150	CDA—under negotiation
First Year Algebra with Applications Project	U. Chicago	36		Not available
Human Behavior Curriculum Project	APA	700		Under preparation
Human Sciences Program	BSCS (U. Colorado)	1,300	500	In testing phase
Individual-ized Science Instruc-tional System (ISIS)	Florida State	3,400	160	Xerox (Ginn)
Mathematics Problem Solving Project	Indiana U.	265		Under preparation

Appendix 1. (continued)

Curriculum Project	Grantee	Develop-ment	Implemen-tation	Publisher, Date of Publishing Agreement
Mathematical Resources Project	U. Oregon	292		Disseminated by grantee
Outdoor Biology Instruc-tional Strategies (OBIS)	U. California, Berkeley	670	300	Sold at cost by grantee
Project for the Mathematical Development of Children	Florida State U.	470	2	Materials under development
Technology-People-Environment (TPE)	SUNY, Stony Brook	134	163	Learning Realities Inc., Under negotiation
Unified Science and Mathematics for Elementary Schools (USMES)	EDC	2,600	1,300	Under negotiation
Totals		101,207	78,581	

Source: NSF Science Curriculum Review Team, *Pre-College Science Currciulum Activities of the NSF, II*, Appendix, May 1975. Adapted from Appendix 7.

Appendix 2
Public Knowledge of Science: Report of a Survey by the National Assessment of Educational Progress

To what extent has the increased dissemination of sophisticated materials in science changed the level of knowledge about science and the understanding of its methods and concepts? The most thorough effort to assess public understanding of science in the United States was conducted by the National Assessment of Educational Progress (NAEP) in a sequence of surveys in 1969–1970 and 1972–1973. The surveys were administered to a national sample of students, ages nine, thirteen, and seventeen, and to a fourth group of young adults, aged twenty-four to thirty-five. They were intended to test the implementation of educational objectives, students' knowledge of fundamental facts and principles of science, their ability to engage in science, their understanding of its investigative nature, and their attitudes. The survey developers assumed that appropriately educated students would understand that science depends on observation and experiment directed intelligently within a logical theoretical framework. They would understand the role of theory in the process of analyzing observations and in making predictions.

The findings of the survey suggest that facts were far better assimilated and recalled than an understanding of the character of science, that students showed only limited understanding of scientific method and were confused about the function of scientific models, theories, hypotheses, and facts. For example, more than half of the respondents in each age group failed to answer correctly questions probing the differences among facts, theories, models, and empirical laws. Given a list of statements, fewer than half of the seventeen-year-olds could correctly differentiate a description of a model from a list of empirical observations. Only 58.7 percent of the seventeen-year-olds and 45.3 percent of the adults tested in 1973 appeared to understand the difference between facts and hypotheses with sufficient subtlety to select the correct answers among multiple-choice items. In one question, students were asked what a scientist might *not* do when beginning a scientific problem (the correct answer being that a scientist would not list the conclusions to be proved). In 1973, only 25 percent of the seventeen-year-olds responded correctly; this represented a 5.8 percent decline in the number of correct answers since the previous survey in 1969.

A number of questions dealt directly with the understanding of evolution theory. One classic question called for explaining the length of the giraffe's neck in terms of the theory of natural selection; 57.1 percent of seventeen-year-olds selected the correct answer, but a statistician, calculating the likelihood of guesses for this question, estimates that probably no more than 40 percent really understood the response. "Even the most optimistic interpretation . . . indicates that 40–50 percent of the nation's seventeen-year-olds did not give evidence of understanding this basic idea of western science . . . a concept [that] requires almost no technical knowledge to understand."[2] Answers to other questions pertaining to evolution also demonstrated considerable ignorance. For example, 68 percent of seventeen-year-olds and 63 percent of adults associated the idea of natural selection with Darwin, but in a more sophisticated question ("How could a fossil of an ocean fish be found on a mountain?") only 39 percent of the adults chose the correct answer that the mountain was raised after the fish was dead. Most respondents believed that the fossil was carried by a flood to the mountain. And in a question about how long man has lived on the earth, 24.6 percent of adults estimated less than 100,000 years and 21.3 percent claimed not to know.[3]

Regional differences in the responses to questions on evolution suggest the possible influence of religious values as well as overall differences in educational quality. Students from the southeastern United States scored about 5 percent lower in the entire survey than respondents from the country as a whole, but they were especially weak in biology. In one exercise concerning what scientists learn from studying fossils, there was a striking 23 percent difference between the southeast and the rest of the nation in the number of correct responses.

In the four years between the two surveys (1968–1969 to 1972–1973), performance in the science survey declined by about 2 percent, interpreted by the NAEP as "significant" and well outside the margin of error, corresponding to a loss of six months "learning experience." Scores declined most systematically in questions relating to the nature of the scientific process. Averaging all the questions in this area, we find that the number of correct answers decreased by about 6 percent among both seventeen-year-olds and young adults.

Notes

1. The NAEP is part of the Education Commission of the States, a consortium of state education officials. Funded by the Office of Education, its offices are in Denver, Colorado. The survey covered 90,000 students and is described in NAEP reports.

2. NAEP, "Gilberts Discusses Meaning of Science Results," *Newsletter* March–April 1974, pp. 6–7.

3. BBC held a similar survey in 1958. Two thirds of their sample had some knowledge of the concept of evolution and defined it in terms of change; one third could volunteer no information at all. Only one in three could associate the theory with Darwin. And in a question about why giraffes have long necks, 50 percent of the response indicated belief in special creation theory, 16 percent in Lamarckian theory, and 33 percent in Darwinian evolution theory. The study concluded that the average viewer "believes in evolution which for him means not much more than that man has descended from monkeys." Described in Stuart Blume, *Toward a Political Sociology of Science*, New York: The Free Press, 1974, 255–256.

Appendix 3
Proposed Creationist Revisions of the California
Science Framework for 1976.

The following document is from a position paper submitted by the
Creation Science Research Center to the California Curriculum Development and Supplemental Materials Commission in October
1975. Proposals for changes in the 1976 guidelines were made by
crossing out statements in the original *Science Framework* and by un-
derlining statements creationists wished to add. The few statements
neither crossed out nor underlined were regarded as acceptable.

Science Framework Changes

~~Another example of the interdependence of the structure and function is found in evolution. Cause and effect evolutionary theories
were at first misinterpreted by Lamarck when he predicted that
function gave rise to structural adaptations. Experimental research
indicates that structures evolved that made some organisms more
adaptable to their environment than others. Organisms that evolved
parts that did not successfully function within their environment did
not survive.~~ The interdependence of structure and function is also
believed to be demonstrated in evolution. Lamarck proposed a
cause-and-effect relationship between the function of the bodily
parts of living organisms and the appearance of structural adaptations in their offspring. Ecological studies suggest that some organisms tend to adapt to changes in their environment better than do
other organisms. Those that are less successful in adapting do not
survive. In Lamarck's theory the need, use, or disuse of a given capacity or organ caused not merely adaptation but the evolution of
new organs and structures in succeeding generations. This theory is
yet to be supported by experimental evidence.

Another order of interactions is that of supposed evolutionary
events, which are believed to have produced ~~produce~~ predictable
changes in certain kinds of objects over long periods of time. One
theory claims that atoms, interacting with one another and evolving

over eons of time, gave rise to the present assemblage of various kinds of elements. Another evolutionary thesis describes the progress of stars all the way from young gaseous nebulae to pulsating dying stars. ~~There is much continuing debate among scientists concerning evolutionary theories of cosmogenesis because of the numerous theoretical problems which remain unsolved and because the postulated developmental process cannot be reproduced experimentally.~~ Still another interacting series of events has produced the ~~evolution~~ <u>transformation</u> of rocks from igneous to sedimentary and matamorphic.

~~Interactions between organisms and their environments produce changes in both.~~ Changes in ~~the~~ <u>earth</u> environment are readily demonstrable on a short-term basis; i.e., over the period of recorded history (circa 5,000 years). ~~These~~ <u>Such</u> changes <u>also</u> have been inferred from geologic evidence over a greatly extended period of time (billions of years). ~~although the further back we go, the less certain we can be. Prehistoric processes were not observed, and replication is difficult.~~ <u>Interaction between populations of organisms and their environments produced changes in both.</u> During the past century and a half, the earth's crust and the fossils preserved in it have been studied intensively by scientists. Fossil evidence shows that <u>many</u> organisms populating the earth have not always been structurally the same. The differences are ~~consistent with~~ <u>interpreted in terms of</u> the theory that anatomical changes have taken place through time. The Darwinian theory of organic evolution postulates a genetic basis for the biological development of complex forms of life in the past and present and the changes ~~noted~~ <u>inferred</u> through time. <u>This theory, since it deals with postulated prehistoric events and processes which neither were observed by man nor can be repeated and controlled by man, is not subject to possible falsification by experimental test, a normal requirement for scientific theories.</u>

The concepts that are the basic foundation for this theory are (1) that inheritable variations exist among members of a population of like organisms; and (2) that differential successful reproduction (i.e., survival) is occasioned by the composite of environmental factors impinging generation after generation upon the population. The theory is used to explain the many similarities and differences that exist between diverse kinds of organisms, living and extinct. The actual variations in living populations observed experimentally and in the field during the decades of modern biological research appear to be limited by rigid genetic boundaries. The theory, therefore, involves the idea of large-scale evolution by the extrapolated accumulation over vast periods of geologic time of limited variations of the type which have actually been observed.

~~The theory of organic evolution, its limitations not withstanding, provides a structural framework upon which many seemingly unrelated observations can be brought into more meaningful relationships.~~ Biologists also have developed, from experiments and observations, hypotheses concerning the origination of life from nonliving matter (e.g., the heterotroph hypothesis). These ideas together with the Darwinian theory in its modern form constitute the explanation for the origin and development of life based upon a materialistic interpretation of the data from pertinent sciences. The theory of organic evolution, its limitations notwithstanding, provides a structural framework within which many seemingly unrelated observations can be brought into more meaningful relationships. There are, nevertheless, data from the biological and physical sciences which are difficult to fit into the materialistic and evolutionary framework of interpretation which has been adopted by a majority of scientists over the past century since Darwin's time.

Another interpretative framework has been adopted by that minority of scientists who hold a theistic and creationist rather than a materialistic and evolutionary philosophy. They believe that the data of the sciences can better be understood and the origin of life better explained in terms of a theory of divine creation in accord with an intelligent, purposeful plan. This theory, in common with the evolutionary theory, is not subject to experimental falsification because it contains postulated events and processes which neither were observed by nor can be repeated by man. Also, there are data from the sciences which are difficult for the creationist to fit into his framework.

~~Philosophic and religious considerations pertaining to the origin, meaning, and values of life are not within the realm of science, because they cannot be analyzed or measured by the present methods~~

of science.

In view of the fact that the theistic and materialistic philosophies are equally religious and/or anti-religious and because both the evolutionary and creationist theories of the origin and development of life are equally inaccessible to falsification by experimental test, both should be studied in the light of the scientific data. The essential philosophy underlying each interpretive system should be clearly identified, but these should not be the subject of study in the science classroom.

Index

Ackworth, Captain, 71
Agassiz, Louis, 10, 11
Alaska, 108
America United for the Separation of
    Church and State, Inc., 51
American Anthropological Association,
    87, 88
American Association for the Advance-
    ment of Science (AAAS), 87
American Civil Liberties Union, 14
American Council of Learned Soci-
    eties, 30
American Institute of Biological Sci-
    ences (AIBS), 27, 28
American Nuclear Society, 131
American Scientific Affiliation (ASA)
    65–66, 76, 83, 95
Archibald, George, 111
Arizona, 51, 59, 107, 108, 111, 116,
    140
Arkansas, 17, 19, 70
    State Supreme Court, 17
Armstrong, Herbert W., 43
Ashford, John, 111
Astrology, 127, 150

Barnes, T. G., 48
Bauman Amendment, 115
Beecher, Henry Ward, 12
Bestor, Arthur, 41
Bible Science Association, 68, 70
Biological Sciences Curriculum Study
    (BSCS), 24, 27–30, 47, 54, 72,
    93, 94, 110, 134, 136
Bob Jones University, 70
Borman, Frank, 77
Bruner, Jerome, 31, 33, 37, 109
Bryan, William Jennings, 14–15, 86
Bube, Richard, 95

California, 1, 3–4, 46, 51–53, 61,
    66–68, 70, 81, 82, 84, 86–88, 93,
    94, 96, 97, 101, 108, 110, 122,
    128, 132, 140, 148
    Board of Education, 67, 81, 82, 100,
        131
    Curriculum Development and Sup-
        plemental Materials Commis-
        sion, 84, 95
    Guidelines, 81, 82
    Orange County, 81
    Science Framework, 82–84, 87, 88,
        95, 96, 100, 128
    State Advisory Committee on Sci-
        ence Education, 82, 84, 85
    State Education Code, 95
    State Textbook Commission, 52
Carnegie, Andrew, 12
Catastrophism, 9, 11
Christian Heritage College, 68
Citizens for Decency Through Law, 48
Citizens for Scientific Creation, 97
Civil Rights Act, 1964, 82
Colorado, 43
Conlan, John, 111–116, 119–120, 123,
    129
Connally, John, 29
Coors, Joseph, 49, 58
Council for Basic Education (CBE), 41,
    49, 112
Creation Research Science Education
    Foundation, Inc., 52, 71
Creation Research Society (CRS), 66,
    67, 72, 81
Creation Science Research Center
    (CSRC), 67–69, 77
Creation theory, 1, 17, 52–54, 60–65,
    67, 69, 70, 72, 82–87, 89, 90,
    92, 98–100, 129, 134
Curriculum Development Associates
    (CDA), 34, 122
Cuvier, George, 9

Darrow, Clarence, 14–15, 86

Darwin, Charles, 9–11, 16, 18, 27, 85, 92, 144, 147, 165
  Darwinian evolution, 5, 50, 166
  Darwinism, 10–12, 86
  Origin of Species, 9, 10
Day, Howard, 83
Deward, Douglas, 71

Education Development Center (EDC), 31, 33, 34, 108, 116, 122
Educational Research Analysts (ERA), 48, 71
Educational Services Institute, 31
Epperson, Susanne, 17
Evolution Protest Movement, 71, 77
Evolution theory, 1, 2, 9–11, 13–17, 22, 27, 29, 30, 52, 53, 57, 60, 62–66, 68, 70, 71, 73, 83, 88–90, 93–96, 98–100, 104, 110, 129, 135, 144, 166

Faubus, Orval E., 17
Federal Communications Commission, Fairness Doctrine, 135
Fischer, Robert, 95
Fleming, J. Ambrose, 71
Florida, 29, 70, 107, 108, 117
Fluoridation, 5, 127, 131, 133
Ford, John, 83, 84, 95, 97
Fosdick, Harry Emerson, 14
Fundamentalism, 13–16, 29, 33, 41–43, 46, 55, 127, 149–151

Gabler, Mel (Mr. and Mrs.), 47–48, 110, 132
Geertz, Clifford, 151
General Accounting Office (GAO), 119, 120
Genesis School of Graduate Studies, 70
Geology, 9, 12, 27, 117
Georgia, 51, 59
Gillispie, Charles, 10
Gish, Duane, 69, 72
Gosse, P. H., 12
Gray, Asa, 10
Grebe, John J., 48
Grose, Vernon, 77, 83, 84, 101, 128, 140

Handlin, Oscar, 145
Hardin, Garrett, 96
Hargis, Billy James, 19, 43
Harward, Thomas, 83
Hays, Arthur Garfield, 15, 61
Helms, Jesse, 119–120
Heritage Foundation, 49, 54, 55, 111, 137
Holt, Marjorie, 111
Hopper, Barbara, 96
Hubbard, David, 83, 95
Hutton, James, 9
Huxley, Julian, 13

Idaho, 70, 108
Individualized Science Instruction Systems (ISIS), 117, 120
Institute for Creation Research (ICR), 68, 69, 71, 72, 74, 75, 134
Irwin, James, 77

Jehovah's Witnesses, 15, 43, 45, 58, 131
  Watchtower Society, 44
John Birch Society, 48, 106

Kentucky, 51
Kilpatrick, James, 110, 114
Kumamoto, Junji, 84

LaHaye, Tom, 68
Lamarck, Jean-Baptiste de, 9, 166, 167
Lammertz, Walter, 81
Leadership Action, Inc., 49, 110, 112
LeClercq, Frederic, 51
Lemmons, Reuel, 29, 30
Lester, Lane, 70, 72
LeTourneau, Richard, 48
Lyell, Charles, 9, 61

McCarthy, Joseph, 53
McGraw, Ona Lee, 132
McIntire, Carl, 43
McKenna, James, 137
Man: A Course of Study (MACOS), 1, 30–35, 37, 48, 104–119, 130, 132, 136, 137, 142, 150
Maryland, 108

Massachusetts, Cambridge, 22, 31
Mayer, William, 93
Merton, Robert, 145
Michigan, 51, 59, 66
Michigan State University, 70
Mississippi, 17, 19
Missouri, 116
Mitchell, Edward, 77
Monod, Jacques, 8
Moore, John A., 85
Moore, John N., 70, 85
Morris, Henry, 69, 71, 134
Morris, John, 69
Mosher, Charles A., 113
Moudy, T. M., 118, 142, 143
    Moudy Committee, 119, 121, 132,
        137
Muller, Hermann J., 16

Nägeli, Karl, 10
National Academy of Sciences (NAS),
    87
National Aeronautics and Space Ad-
    ministration (NASA), 73
National Assessment of Educational
    Progress (NAEP), 147, 164
National Association of Biology Teach-
    ers (NABT), 51, 61, 86, 87,
    92–94
National Coalition for Children, 110,
    132
National Defense Education Act, 36,
    141
National Education Association (NEA),
    13, 132
National Science Foundation, 1, 2, 4,
    22–28, 31, 34–36, 42, 49, 54,
    101, 112–121, 132, 136
    Division of Social Sciences, 26
    National Science Board, 118, 119,
        121
    Science Curriculum Review Team,
        25, 35
National Science Teacher's Association,
    17, 143
Natural selection, 10–12, 16, 60, 61,
    72, 165
Nature Study Movement, 21

Netsilik Eskimos, 31–33, 109, 110, 130
New Mexico, 29
Newton Scientific Organization, The
    (Great Britain), 71
New York, 108
Nixon, Richard M., 120
Nuclear Power, 5, 127, 131

Ohio, 52, 71
Oregon, 108
    School board, 52
Owen, Richard, 11

Passmore, John, 140
Patriotism, 26, 41, 73, 137
Pennsylvania, 108
Physical Science Study Committee
    (PSSC), 22
Piel, Gerard, 118
Progressive Movement, 21, 49
Proxmire, William, 113

Rafferty, Max, 81–83
Ragle, Eugene, 83
Reagan, Ronald, 101
Rickover, Hyman, 21
Riecken, Henry W., 26
Roberts, Oral, 43
Rockefeller, John D., 12

Schlei, Norbert A., 81
Scientific American, 118
Scientific Creationism Association of
    Southern New Jersey, 71
Scientific Secularism. *See* Secular Hu-
    manism
Scopes Trial, 3, 14–17, 84
Scott, Gary, 17
Secular Humanism, 4, 44, 54–57, 76,
    81, 82, 90, 109, 110, 130, 137,
    141
Sedgwick, Adam, 11
Segraves, Nell, 81
Seventh Day Adventists, 15, 44, 45, 96
Sex Education, 16, 26, 117
Shaw, George Bernard, 40
Smith, D. O., 19
Southern Illinois University, 70

Spencer, Herbert, 12
Steiger, Samuel, 116
Steinbacher, John, 110, 122
Stever, H. Guyford, 54, 113, 115, 120, 121
Sumrall, Jean, 81
Symington, James W., 113, 116

Teague, Olin, 113, 116, 118, 132
Tennessee, 14, 17, 50–51, 59, 61, 108
  Court of Appeals, 51
  Dayton, 15, 50, 61, 85
  General Assembly, 50
  House of Representatives, 51
  Jacksboro, 17
  Nashville, 51
Texas, 3, 19, 29, 46, 48, 51–52, 71, 108, 110, 118, 128, 132, 140
  Education Policy Act, 48
  State Textbook Depository, 30
Texas Christian University, 118
Textbook publishers, 16, 21, 22, 26, 33, 34, 36, 51, 82, 84, 94, 96, 114, 144
  Follett, 114
  Lippincott, 114

UFO cults, 128
Uniformitarianism, 9, 11, 62
United Nations, 41
United States Congress, 1, 26
United States Constitution
  First Amendment, 14, 30, 51, 54, 90
  Fourteenth Amendment, 51, 90
United States House of Representatives, Committee on Science and Technology, 111, 113, 116, 119, 121, 141
United States Senate, Committee on Labor and Public Welfare, 113
United States Supreme Court, 17, 42, 48, 51, 54, 81, 133
University of California, 72, 88
University of Colorado, 28, 54
Ussher, James, 9

Velikovsky, I., 42, 140
Vermont, 108

Virginia Polytechnic Institute, 71
Von Braun, Wernher, 73, 77
Von Daniken, E., 42, 140

Waddington, C. H., 13
Wakely, Cecil, 71
Washington, 51
Washington, D.C., 49, 54, 92, 110, 111
Washington Evening Star, 1, 54
Weber, George, 49
Welch, Claude, 93
West Virginia, 55
  County Board of Education, 54, 56
  Kanawha, 54–56, 113, 132–133, 142, 149
Willoughby, William, 54, 132, 136
Wisconsin, 53

Zacharias, Jerrold, 22